149

Arachnida

FRONTISPIECE: *Ricinoides afzelii*
Photograph taken by Michael Tweedie

Arachnida

THEODORE SAVORY

2nd edition

Est locus, in geminos ubi bracchia concavat arcus
Scorpios, et cauda flexisque utrimque lacertis
Porrigit in spatium signorum membra duorum
Ovid

1977

ACADEMIC PRESS

London · New York · San Francisco

A Subsidiary of Harcourt Brace Jovanovich, Publishers

ACADEMIC PRESS INC. (LONDON) LTD
24–28 Oval Road,
London NW1

U.S. Edition published by
ACADEMIC PRESS INC.
111 Fifth Avenue,
New York, New York 10003

Library of Congress Catalog Card Number: 76-1099

ISBN: 0-12-619660-5

Text set in 11/12 pt. Monotype Baskerville, printed and bound
in Great Britain by The Garden City Press Limited, Letchworth, Hertfordshire SG6 1JS

Preface

The stimulus to attempt a revision of this book arose from my attendance at the Sixth International Arachnological Congress at Amsterdam in 1974. I was fortunate to be able to communicate my inspiration to Academic Press, to whom my gratitude must be manifest.

Several changes have seemed to be desirable, owing to the rapid advances in Arachnology during the past ten years, and I have dropped some sections in order to expand others. I have taken the long overdue step of promoting the Cyphophthalmi to higher rank, and have changed the former sequence of the other orders to bring them into closer representation of their probable phylogeny. I have given them the names that I tried to justify in *Systematic Zoology* in 1972. Since "Spiders, Men and Scorpions" is now out of print I have restored the historical chapter to narrative form.

In addition, Chapter 26 on the Acari has been written for me by Keith R. Snow and Chapter 30 on Economic Arachnology has been revised by Martin H. Muma. I am most grateful to these authors for thus helping me to conceal some of my ignorance. Further, I wish to thank the following for permission to make use of figures published by them: D. J. Curtis for Fig. 2, R. R. Jackson for Fig. 107, C. Juberthie for Figs 71–74, Max Vachon for Fig. 19, and Messrs Routledge and Kegan Paul for Figs 80 and 81.

Every reader will notice my indebtedness to my contemporaries, who are carrying out the researches that propagate our science. They are the true creators of this book, of which I have been the fortunate compiler, co-ordinator and interpreter, acting as the mouthpiece through which they have spoken. Their names are recorded below: their help, and often their friendship, have made my work an uninterrupted pleasure.

T.H.S.

Dorking, Surrey
1977

Contents

Preface v
Acknowledgements ix
Outline of Classification xi

I. PROLEGOMENA

1. The Phylum Arthropoda 3
2. The Class Arachnida 7

II. DE ARACHNIDIS

3. Morphology: External Appearance 17
4. Physiology: Internal Organs 31
5. Embryology: Development 44
6. Ontogeny: Growth 49
7. Bionomics: General Habits 60
8. Ethology: Behaviour 71
9. Zoogeography: Distribution 83
10. Ecology: Migration and Dispersal 87
11. Phylogeny: Evolution 94
12. Taxonomy: Classification 103

III. PROLES ARACHNES

13. The Order Scorpiones 115
14. The Order Palpigradi 125
15. The Order Uropygi 132
16. The Order Schizomida 138
17. The Order Amblypygi 143
18. The Order Araneae 148
19. The Order Kustarachnae 168
20. The Order Trigonotarbi 170
21. The Order Anthracomarti 173
22. The Order Haptopoda 175
23. The Order Architarbi 177
24. The Order Opiliones 180

25. The Order Cyphophthalmi 191
26. The Order Acari 198
27. The Order Ricinulei 211
28. The Order Pseudoscorpiones 220
29. The Order Solifugae 233

IV. DE ARACHNOLOGIA

30. Economic Arachnology 247
31. Historical Arachnology 258
32. Practical Arachnology 270
33. Chemical Arachnology 282
34. Medical Arachnology 288
35. Linguistic Arachnology 295

V. HETEROGRAPHIA ARACHNOLOGICA

36. The Spider's Web 303
37. Courtship in Arachnida 309
38. Arachnophobia 316
39. Arachnida in Amber 320

VI. EPILEGOMENA

Bibliography 325
Index Rerum 335
Index Animalium 338

Acknowledgements

The general terms that I have used in my Preface must be expanded to make specific mention of those to whom I owe so much. They fall into three groups.

In the first list are those to whom I listened and with whom, through the lips of a multilingual wife, I talked at Amsterdam in 1974. They were: P. Bonnet, J. A. L. Cooke, C. D. Dondale, E. A. G. Duffey, P. D. Gabbutt, H. Homann, O. Kraus, B. H. Lamoral, P. T. Lehtinen, R. Legendre, H. W. Levi, A. F. Millidge, W. B. Muchmore, J. R. Parker, N. Platnick, M. Rambla, G. S. Rovner, M. Vachon, L. Vlijm, B. R. Vogel and P. N. Witt.

The second list names those to whom I have since written and who have sent me copies of their publications. In addition to those mentioned above, information of many kinds has come to me from: S. C. Anderson, T. S. Briggs, J. L. Cloudsley-Thompson, B. Condé, A. L. Edgar, R. R. Forster, C. J. and M. L. Goodnight, V. V. Hickman, C. Juberthie, R. Leech, Z. Maretic, B. J. Marples, E. A. Seyfarth, L. Størmer and P. Weygoldt.

Finally, I thank my former pupils, who have been ready, as always, to help me with their special knowledge in various difficulties: A. Coyle, E. A. Eason, D. L. Harrison, J. Pollock, A. S. C. Ross and J. C. A. Raison. Happy the schoolmaster whose invested efforts pay such dividends as these after so long an interval of time.

MAIAE CARISSIMAE
UXORI SECUNDAE
LIBRUM MEUM
QUADRAGENSIMUM
AMORIS PLENUS
D.D.D.

Outline of Classification
adopted in this book

Phylum Arthropoda
  Sub-phylum Trilobitomorpha
    Class Trilobita
  Sub-phylum Pycnogonida
    Class Pantopoda

  Sub-phylum Chelicerata
    Class Merostomata
      Sub-class Xiphosura
      Sub-class Eurypterida
    Class Arachnida
      Sub-class Scorpionoidea
        Order Scorpiones
      Sub-class Arachnoidea
        Order Palpigradi
        Order Uropygi
        Order Schizomida
        Order Amblypygi
        Order Araneae
        Order Kustarachnae
      Sub-class Opilionoidea
        Order Trigonotarbi
        Order Anthracomarti
        Order Haptopoda
        Order Architarbi
        Order Opiliones
        Order Cyphophthalmi
        Order Acari
        Order Ricinulei
      Sub-class Chelonethoidea
        Order Pseudoscorpiones
        Order Solifugae

  Sub-phylum Mandibulata
    Class Crustacea
    Class Insecta (Hexapoda)
    Class Chilopoda
    Class Diplopoda

I. PROLEGOMENA

1

The Phylum Arthropoda

The phylum Arthropoda von Siebold and Stannius, 1845, containing as it does, some 700,000 species, includes about four-fifths of all the known animals of the world: both in the number of species and the number of individuals it far surpasses all the other phyla together.

Its members are characterized by the possession of a metamerically segmented body, protected by an exoskeleton of chitin. The segments, or somites, are united by a softer pleural membrane, and are grouped in sets, or tagmata, such as are to be seen in the head, thorax and abdomen of an insect. In addition to the somites there is an acron anteriorly and a telson posteriorly.

Each somite typically carries a pair of appendages, hollow outgrowths of the ectoderm, provided with joints, from which the name of the phylum is taken. At least one pair of these appendages is modified to act as mouth parts.

The general body cavity is haemocoelic and the coelom is much reduced. Coelomoducts are present as excretory organs and as gonoducts.

The central nervous system, consisting of cerebral ganglia followed by a solid, ventral, ganglionated nerve cord, is strongly reminiscent of the annelidan nervous system, and supports belief in the evolution of arthropods from annelidan stock.

There are no cilia and no nephridia.

The sexes are usually separate and the eggs are rich in yolk. Growth is discontinuous and is associated with ecdysis and often with metamorphosis. Rhythmic and instinctive types of behaviour are highly developed, and a social organization is not unknown.

Within the wide boundaries of this diagnosis, a great variety of forms, both living and fossil, have been discovered and described. So great are the diversities that it has been suggested that the Arthropoda have had not one but several centres of origin among the ancestral annelids, and that the arthropods existing today are the surviving representatives of a number of lines of parallel evolution.

Present conventional classification divides the most successful groups into a crustacean–myriapod–insect moiety and a eurypterid–arachnid–pycnogonid moiety, which seem to be reasonably distinguished as the sub-phyla Antennata (or Mandibulata) and Chelicerata. This implies that they are the descendants of a common ancestor, while somewhere on the fringe, as it were, stand the Onychophora, the Tardigrada or water-bears, and the Pentastomida or Linguatulida. The relationships between these three groups and the rest are not obvious.

The Onychophora, which include a score or so of species of Peripatus and its allies, were at one time thought to form a link between the Annelida and the Arthropoda, both of which they resemble to some extent. There is, however, reason to believe that they are neither the descendants of the Annelida nor the ancestors of the Arthropoda. Less easily placed in a simple and satisfying scheme of classification are the Tardigrada and Pentastomida.

The opinions of Manton (1973) can never be neglected. After an exhaustive study of the basic structure of the limbs and mandibles, combined with precise observation of the actions involved in moving and feeding, she can avoid the risk of assuming "functionally impossible ancestral stages" and suggests that the Arthropoda be arranged in three phyla. These are the Crustacea, the Chelicerata and the Uniramia. In the last, which includes the Onychophora, the Myriapoda and the Hexapoda, the limbs are fundamentally uniramous, contrasting in this respect with the basically biramous limbs of the Crustacea and Chelicerata.

Complete acceptance of these views would result in the following tabular summary:

Super-phylum Arthropoda
   Phylum Trilobita
   Phylum Crustacea
   Phylum Uniramia
      Class Onychophora
      Class Myriapoda
      Class Hexapoda
   Phylum Chelicerata
      Class Pycnogonida
      Class Eurypterida
      Class Xiphosura
      Class Arachnida

The Chelicerata, with which alone this book is concerned, may at this point be characterized as follows:

Arthropoda with a prosoma of six somites and an opisthosoma of 12 somites. The acron is not separable from the first somite, which carries one pair of pre-oral appendages, the chelicerae, primitively chelate organs. There are five pairs of post-oral appendages, the first of which, the pedipalpi, may be indistinguishable from the legs which follow it, or may be specialized for certain functions.

In the opisthosoma the first somite may be reduced or may be absent. The genital orifice always opens on the second somite. The sexes are always separate. There is, in some orders, a visible distinction between a mesosoma of seven and a metasoma of five somites.

Chelicerata may be aquatic, terrestrial or secondarily aquatic, respiration being by gills, book-lungs or tracheae.

The nature of the relationship between the Chelicerata and the other groups of Arthropoda has long been difficult to define. The suggestions put forward in the past, though based as they have been on careful research and considered opinion, have differed in an extraordinary way from one another.

At one extreme stands the early theory of Savigny (1816), that Arthropoda may be looked upon as headless Crustacea, that is to say that from a common ancestor one class has retained, while the other has lost, the anterior or cephalic somites. This implies that the chelicerae and pedipalpi of Arachnida are modified legs. At the other extreme are the hypotheses of Börner (1921) and Henriksen (1918), who regarded the first three pairs of legs of Arachnida as the mandibles and maxillae of Mandibulata, and so picture Arachnida as walking on their mouth parts.

The consensus of modern opinion depends on the apparent homology between the chelicerae of Chelicerata and the second antennae of Mandibulata. A comparison between the chief classes of Arthropoda then takes the following form:

MEROSTOMATA	ARACHNIDA	CRUSTACEA	INSECTA
1.		Antennae i	Antennae
2. Chelicerae	Chelicerae	Antennae ii	—
3. Legs i	Pedipalpi	Mandibles	Mandibles
4. Legs ii	Legs i	Maxillae i	Maxillae
5. Legs iii	Legs ii	Maxillae ii	Labium
6. Legs iv	Legs iii	Maxillepedes i	Legs i
7. Legs v	Legs iv	Maxillipedes ii	Legs ii
			Legs iii

Too much significance should not be attached to the above table, which is founded on the assumption of comparable rates and methods of evolution in the classes concerned, while the bulk of the evidence is against rather than in favour of such a correspondence.

The comparison implied loses much of its value if the Arthropoda are not regarded as a monophyletic group. As the evidence for its polyphyletic nature becomes steadily more convincing, the possibility of accurately defining analogies between all the somites and appendages of the head and thorax becomes proportionately more difficult.

This leads to a further point. In the body of an insect there is a clear distinction between the head and the thorax, the latter name being given to the three somites that carry the three pairs of legs. The head of an arachnid is not so easily defined. A reasonable view is that not more than one somite can be called cephalic, and its distinction from the thorax is of little significance.

2
The Class Arachnida

Chelicerate Arthropoda, in which the adult body is fundamentally composed of 18 somites, divisible into a prosoma of six and an opisthosoma of 12 units. Segmentation may be obscured in either or both of these tagmata by the fusion or suppression of the sclerites. There are not more than 12 simple ocelli. The prosoma carries six pairs of appendages, the chelicerae, the pedipalpi and four pairs of legs. The chelicerae, in front of the mouth, are of two or three pieces and are either chelate or unchelate. The pedipalpi are of six podomeres, are either chelate weapons or leg-like limbs, and often carry gnathobases on their coxae. The legs are of seven podomeres: the anterior pair or pairs may carry gnathobases, the tarsi end in two or three smooth or pectinate claws. The sternum may be visibly segmented or entire, and is often reduced, hidden or obliterated by the encroachment of the coxae. The opisthosoma may be united to the prosoma by a narrow pedicel, the seventh somite, and is usually without appendages. The respiratory system, consisting of book-lungs or tracheae or both, open on this part of the body. The sexes are separate, with orifices on the lower side of the second opisthosomatic somite. Courtship is usual and is often elaborate. The individuals are generally terrestrial, carnivorous, nocturnal and cryptozoic. Instinctive behaviour is highly developed and social organization is very rare.

Carolus Linnaeus, in "Systema Naturae", had given inadequate attention to the invertebrates, and improvements to this part of his great achievement could not long be delayed. The first biologist to undertake this task was Lamarck.

Jean Baptiste de Monet, Chevalier de Lamarck (1744–1829), botanist at the Jardin du Roi in Paris, was elected at the age of 50 to be the professor in charge of "Insectes et Vermes" at the newly-founded Muséum d'Histoire Naturelle, becoming thereby the first occupant of a chair of invertebrate zoology. To him fell the task of creating a new system of classification, which he performed with such genius that much of his work has survived until today.

The Linnaean Insecta were exactly the same as the phylum Arthropoda, as defined in 1845 by C. T. E. von Siebold and H. Stannius, and they therefore included a number of wingless forms, which were united in a common order, the Aptera. The true Insecta, with wings and six legs, were placed by Lamarck in a class Hexapoda, but this name has not been universally adopted, that of the parent group, Insecta, being usually retained.

From the Aptera Lamarck extracted several classes, one of which was "Classe Troisième—les Arachnides". At that time it contained scorpions, spiders and mites, together with some excusable intrusions. The class Arachnida, as here recognized, contains 12 living and five extinct orders:

ACARI	†HAPTOPODA	SCHIZOMIDA
AMBLYPYGI	†KUSTARACHNAE	SCORPIONES
†ANTHRACOMARTI	OPILIONES	SOLIFUGAE
ARANEAE	PALPIGRADI	†TRIGONOTARBI
†ARCHITARBI	PSEUDOSCORPIONES	UROPYGI
CYPHOPHTHALMI	RICINULEI	

† Extinct orders

The order named Pedipalpi by Latreille was clearly heterogeneous. It was divided by Millot in 1942 into three orders: Phrynides, Thelyphonides and Tartarides, and independently in 1945 by Petrunkevitch into the order Phrynichida, Thelyphonida and Schizomida. This introduction of new names for groups which already had familiar names in their status as sub-orders was a needless complication, and later writers have wisely used the older names Amblypygi and Uropygi.

An order named Poecilophysida was founded in 1876 by Pickard-Cambridge on several examples of primitive stomatostigmatic mites from Kerguelen's Land: they had been placed by Thorell in 1871 in the genus Rhagidia.

For readers who may be unfamiliar with some of these Arachnida, a short survey of the living orders will form a helpful introduction to the now well-established science of arachnology.

Scorpiones
Scorpions: formidable arachnids, perhaps among the first animals to leave the water and live on the land. Since the dawn of civilization, scorpions, conspicuous with their large pincers and poison-sting, have been feared by man; they have appeared in myths and have been granted a celestial place in the Zodiac.

Palpigradi
Micro whip-scorpions: diminutive arachnids, leading hidden lives in

warm countries and deserving, because of their very primitive structure, more attention than they usually obtain.

Uropygi
Whip-scorpions: flat-bodied arachnids with a long whip-like telson, forming a tropical order now beginning to be better appreciated and more carefully studied.

Schizomida
A small group, related to the above, about which we are not learning very much.

Amblypygi
Tailless whip-scorpions: more flat-bodied arachnids, whose whips are not telsons, but very long and thin first legs. Another order that is beginning to come to the fore.

Araneae
Spiders: dominant arachnids, known to all men as the spinners of silk threads that are often used to make elaborate, ingenious and beautiful webs. There are many species, showing a wide range in structure and habits.

Cyphophthalmi
Primitive arachnids: until now included with the harvestmen, to which they are ancestrally related.

Opiliones
Harvestmen: ludicrous arachnids with two eyes perched on a small body, often bizarre in form, and supported by legs too long for convenience and somewhat insecurely attached. Recognized by most people as clearly different from spiders, there are a score or so of species in Britain, several of which are commonly to be seen in woods and gardens in the autumn.

Acari
Mites and ticks: small arachnids, some of them aquatic and some parasitic. They include the only arachnids of much economic importance and the only ones studied by economic zoologists. The order is less homogeneous than any other and has sometimes been placed in a separate class.

Ricinulei
Mysterious arachnids: scattered over parts of the tropical belt, and at

one time so rare that every fresh specimen was something of a zoological triumph, but now better known and obtainable by the hundred.

Pseudoscorpiones

False scorpions or book-scorpions: fascinating arachnids, like tiny tailless scorpions, the very largest about one-third of an inch long. They are ubiquitous and quite inexplicably neglected by zoologists until recently. Two dozen species occur in Britain, and the sifting of a handful of fallen leaves seldom fails to produce one or two.

Solifugae

Wind-scorpions, camel-spiders or sun-spiders: desert arachnids, amongst the most powerfully armed animals in the world, some with jaws as long as their bodies, some able to climb and run at great speed.

This quick review serves to emphasize the diversities in general familiarity, in distribution and in the sizes of the species that are to be found among the living Arachnida. Two additional comments will illustrate a fundamental problem of theoretical arachnology.

The first of these is the fact that for many of the orders there exists a special and obvious feature, peculiar to one order only, and distinguishing it immediately from all the others. Thus:

Acari	alone have vegetarian and parasitic forms
Araneae	alone have opisthosomatic silk glands
Palpigradi	alone have a projecting proboscis
Pseudoscorpiones	alone have cheliceral silk glands
Ricinulei	alone have a cucullus in front of the carapace
Schizomida	alone have a short telson
Scorpiones	alone have pectines
Solifugae	alone have malleoli
Uropygi	alone secrete acetic acid.

The second comment concerns the sporadic distribution among the orders of special modifications of the basic structure. For example:

(i) The carapace is segmented in Palpigradi, Schizomida and Solifugae.

(ii) No sternum separates the coxae, which meet in the middle line, in Cyphophthalmi, Opiliones, Pseudoscorpiones and Solifugae.

(iii) Pedal tarsi are composed of many pieces on Opiliones, Pseudoscorpiones, Ricinulei and Scorpiones.

(iv) Chelicerae are pointed, not chelate, in Amblypygi, Araneae and Uropygi.

(v) Pedipalpi are chelate in Pseudoscorpiones, Ricinulei and Scorpiones.

(vi) Gnathobases are present on the first pedal coxae in Cyphophthalmi, Opiliones and Scorpiones.

(vii) Offensive fluids are secreted by Cyphophthalmi, Opiliones and Uropygi.

(viii) The first legs are used as tactile organs, rather than for walking in Amblypygi, Palpigradi and Solifugae.

(ix) Trichobothria are absent from Cyphophthalmi, Opiliones, Ricinulei and Solifugae.

(x) A long mobile telson is present in Palpigradi and Uropygi.

(xi) Venom is secreted by Araneae, Pseudoscorpiones and Scorpiones.

(xii) Silk is secreted by Acari, Araneae and Pseudoscorpiones.

A distribution of various important features as irregular as this emphasizes a fundamental fact about the orders of Arachnida; namely that they are all so different from one another that their mutual relations, wherever they exist, are obscured. Attempts to find a path through the resultant maze will be found in Chapters 11 and 12.

The number of species that constitute any order can be given accurately in very few classes of animals, and it is impossible to give more than an estimate of the numbers of Arachnida. For many years what may be called traditional totals were associated with the different orders, and a comparison of the numbers for 1939, a date that chooses itself for any discussion of this kind, and the estimated numbers for 1974, 35 years later, is interesting.

	1939	1974
Acari	6,000	10,000
Amblypygi	60	770
Araneae	50,000	35,000
Cyphophthalmi	20	65
Opiliones	1,600	4,000
Palpigradi	20	50
Pseudoscorpiones	1,000	2,000
Ricinulei	15	35
Schizomida	30	35
Scorpiones	600	750
Solifugae	600	800
Uropygi	70	85

The following considerations apply to both Acari and Araneae, and probably also to the Opiliones.

With the increase in the numbers of interested zoologists and with a consequent increase in the number of collecting expeditions, penetrating into more and more inaccessible regions, the stream of newly discovered species has grown from a trickle to a flood. The time may be not far distant when even full-time specialists will be overwhelmed by the cascade of "new" species. A consequence of this will be a recurrence of the state of affairs that existed in the late nineteenth century: the descriptions become so numerous, the literature so extensive, the languages used so diverse, that the individual systematist is certain to miss records of many species unfamiliar to himself. Names, which will one day be recognized as synonyms, will be given by the hundred before the end of the present century.

The history of the nomenclature of spiders gives support to this prophecy. In Bonnet's "Bibliographia Araneorum" there are references to about 250,000 names, and it is evident that on an average each species has received five names. This reduces an estimate of a quarter of a million species to about 50,000. But in his last volume Bonnet informs us that about 15,000 names have been used once only in all the literature before 1939: it is safe to assume that nearly all of these have been described, and are now known, by other names; and this reduces 50,000 to 35,000. The same is true, if less emphatically, of the names that have been used twice only, so that critical analysis must put the world's spiders known in 1940 at about 30,000.

It is axiomatic that the largest orders will have the widest range, but there is more to the subject of geographical distribution of the Arachnida than this. A simple survey, which will be elaborated in a later chapter, shows that the 12 living orders may be described as follows.

(i) Four orders which are widely dispersed all over the world, including even the frigid regions of the sub-arctic and sub-antarctic: Araneae, Pseudoscorpiones, Acari and Opiliones.

(ii) Four orders which are confined to the tropical or hot subtropical belt, but which are widespread within it: Scorpiones, Amblypygi and Uropygi.

(iii) Four orders which are sporadically distributed and found only in limited and well-separated areas: Cyphophthalmi, Palpigradi, Ricinulei and Schizomida.

A biological feature that follows immediately on the subject of distribution, and is connected with it, is that of size. In general terms, Arachnida are among the small animals of the world, but, by itself "small" is a rather meaningless adjective. The facts may be summarized as follows.

(a) There are four living orders in which the length of the largest species does not exceed 1 cm. These are Pseudoscorpiones, Palpigradi, Schizomida and Cyphophthalmi.

(b) There are four orders in which the length of the largest species exceeds 6 cm. These are Araneae, Uropygi, Solifugae and Scorpiones.

(c) This leaves four orders of intermediate or average size: Acari, Ricinulei, Opiliones and Amblypygi.

It is to be observed that in each group there is an order, named first, which has a wide or very wide distribution, as well as an order, named second, which has a very limited distribution. From this it follows that no correlation between size and distribution can be detected when a whole order is considered as an indivisible unit. Geographical barriers must affect smaller units, sub-orders or families, and will be considered in the appropriate chapters in Part III.

Distribution in space leads to a consideration of distribution in time.

The Eurypterida were aquatic animals, whose fossilized remains have been found in all the Primary strata from the Cambrian to the Permian, but in no later rocks. Probably most of them were marine, but there is evidence that some lived in streams and lakes. The earliest known scorpion is the Silurian species, *Palaeophonus nuncius*; Carboniferous scorpions are fairly numerous: they showed most of the features that characterize living forms and it appears that the order really reached its acme during this era.

The Devonian strata have yielded both spiders and mites, and all remaining orders, except Cyphophthalmi and Pseudoscorpiones, are represented in the Carboniferous, many of them, notably Solifugae and Opiliones, in forms which differed in no essentials from the species living today.

Five orders which rose to comparative prominence during the Carboniferous have since disappeared. The Anthracomarti survived until the Permian, to this extent outliving Kustarachnae, Haptopoda, Architarbi and Trigonotarbi. All these orders have left no more evidence of their ancestors than of their descendants.

These facts are remarkable, and the conclusion to which Berland (1933) has come is sufficiently striking. Such evolution as can be perceived, he writes, is not a progressive change, and progress is the essence of evolution, but a mere replacement of one form by another, apparently equivalent or comparable to it. This implies that the study of fossil Arachnida leads to the conclusion that the hypothesis of an evolution taking place by slow successive degrees is simply not in accordance with the facts.

The subject of the geological record of Arachnida and its bearing on the course of their evolution are considered more fully in Chapter 11.

The most striking features to be mentioned here are the imperfection of the record, with its large gap in the Secondary Era, and the difficulty of arranging the 16 orders in a satisfactory sequence.

From what has been written above it should be clear that external features simplify the placing of any species in its correct order. An elaborate key is therefore a conventional addition, rather than a practical necessity, yet, since it is often expected, a simple form follows. There are several different ways in which such keys may be printed; the method used here, and elsewhere in this book, gives to every descriptive clause a number of its own, and never an asterisk or a number shared with another clause. A number in parentheses is the alternative to the number that precedes it.

1	(2)	Prosoma and opisthosoma joined across whole breadth	3
2	(3)	Prosoma and opisthosoma separated by a pedicel	13
3	(4)	Body divided into proterosoma and hysterosoma between legs 2 and 3	ACARI
4	(3)	Body divided into prosoma and opisthosoma behind legs 4	5
5	(6)	Pedipalpi enlarged, chelate	7
6	(5)	Pedipalpi normal, leg-like	9
7	(8)	Post-abdomen narrow, with terminal sting	SCORPIONES
8	(7)	Post-abdomen normally elliptical	PSEUDOSCORPIONES
9	(10)	Chelicerae large, malleoli on legs	SOLIFUGAE
10	(9)	Chelicerae normal	11
11	(12)	Odoriferous glands opening on carapace surface	OPILIONES
12	(11)	Odoriferous glands opening near tips of tubercles	CYPHOPHTHALMI
13	(14)	With terminal abdominal spinnerets	ARANEAE
14	(13)	No abdominal spinnerets	15
15	(16)	Cucullus at front of carapace	RICINULEI
16	(15)	No such cucullus	17
17	(18)	Telson long, as flagellum	19
18	(17)	Telson short or absent	21
19	(20)	Pedipalpi large and spinous	UROPYGI
20	(19)	Pedipalpi leg-like	PALPIGRADI
21	(22)	Flagellum short, four joints	SCHIZOMIDA
22	(21)	Flagellum absent	AMBLYPYGI

A dichotomic table, written in much greater detail and with extra alternatives, and including also the fossil (extinct) orders, is given by Max Vachon in "Encyclopédie de la Pléiade" (1963), Zoologie, Tome II, pp. 101–5.

II. DE ARACHNIDIS

3

Morphology: External Appearance

The chief external feature of Arachnida is the division of the body into two parts, properly to be called the prosoma and the opisthosoma. The former, composed of a united head and thorax, is often called the cephalothorax; the latter is known as the abdomen. The objection to the use of these terms is that they are also in use for other animals, in which their constitution is not the same as in Arachnida. The prosoma of an arachnid appears to be made up of at least six somites, with the acron indistinguishably fused to the first; and each somite carries a pair of appendages: chelicerae, pedipalpi or legs, which are present in all normal Arachnida without exception. The dorsal plates or tergites of the prosoma are usually fused to form a carapace, which may be uniform and almost free from traces of its segmental origin, but in some orders the posterior tergites are separate. The sternites of the lower surface show a considerable variation.

In the opisthosoma a maximum of 13 somites is recognizable. The first of these may be present only in the embryo and be missing from the adult body. In some orders there is a distinction between a mesosoma of seven and a metasoma of five somites: in others a few posterior somites may form a pygidium. A telson may or may not be present. In general the opisthosoma carries no appendages, the spinnerets of spiders being an obvious and familiar exception.

The subject of the segmentation of the arachnid body is one of some uncertainty and difference of opinion, and it is probable that the account given here may need correction in the future.

At the front, the primitive acron is seldom recognizable. If it persists at all it is so closely fused with the first or cephalic somite that no distinction remains. Belief in a rostral somite behind it is no longer maintained, and it is usual to regard the head as composed of a single somite only, with the chelicerae as its appendages. These are the only appendages in front of the mouth, which during development moves backwards from its primitive anterior position, so that the chelicerae, formerly behind it, move to a forward position. This tendency of the mouth to move back is frequently to be found in the Arthropoda, and is further mentioned in Chapter 10.

The second somite carries the pedipalpi, and somites 3 to 6 carry the four pairs of legs and complete the prosoma.

The seventh somite is of a more diverse character. In Limulus it marks the posterior border of the mouth and carries the chilaria, organs which in the Eurypterida are represented by the metastomatic plate. In the scorpions and the other Arachnida in which the prosoma and opisthosoma are joined across their whole breadth, the somite is not present in the adults, but its temporary existence has been observed in the embryo scorpion. This somite persists in spiders and Ricinulei, and may normally be recognized wherever a pedicel unites the two portions of the body. This narrowing of the somite to form a slender pedicel is the final stage in a process which, like the backward movement of the mouth, is found to different extents in the different orders of Chelicerata. The somite is often called the pregenital somite, since it lies next in front of the genital orifice. The occasional existence of both a tergite and a sternite, the lorum and the plagula, above and below the pedicel, clearly indicates its nature as a separate somite, even when it is much reduced in diameter.

The lower surface of the prosoma presents very different aspects. Fundamentally it should be expected to show a series of six protective sternites, a condition which is in fact found in the extinct genus Stenarthron. In living Arachnida it is modified by the position of the mouth, by the different extents to which the processes of the coxae or gnathobases take a share in the mastication of the food, and by the approach of the coxae of the legs towards the middle. When this is complete, as in Solifugae, there may be no sternites at all, or, as in Opiliones, they may be covered by the coxae and so be normally hidden. Alternatively, there may be a broad and conspicuous sternum derived from four sternites, as in spiders, with the labium in front and the plagula behind.

The opisthosomatic somites behind the pregenital are not more than 11 in number and are to be reckoned as somites 8 to 18.

The eighth somite invariably carries the genital orifice and one or two immediately following it bear the respiratory book-lungs. A distinction between the seven somites of the mesosoma and five somites of the metasoma is obvious in Euypterida and scorpions; but generally the outline is a smooth oval. The hindmost somites provide a very interesting example of comparative morphology.

Hansen and Sørensen pointed out that the last three somites of the opisthosoma show a tendency to form a narrower and semi-independent portion in "Pedipalpi", Palpigradi, Ricinulei and Araneae. In fact, they suggest the inclusive term Arachnida micrura for these orders.

In Uropygi the relative mobility of the tenth to the twelfth opisthosomatic somites may well enable the animal to direct the secretion of its acid-producing glands. In Amblypygi there are no such glands and the "tail" is less obvious. In Palpigradi the mobility of this part is probably a help in the use of the flagellum, whatever this may be. In Araneae, it is supposed that the value of the "tail" is that it separates the anus from the spinnerets and so prevents the latter from being soiled. The habits of Ricinulei are so little known that no purpose can be suggested for the tail in this order. Nevertheless it is undeniable that these four orders do very definitely resemble each other in having a kind of tail, typically formed from the last three somites of the body.

Similar comparisons may be made when the presence of a post-anal telson is considered. In Arachnida post-anal structures are very different in form and function. The scorpions have a short sharp perforate sting with a swollen base enclosing a poison sac. The Uropygi and Palpigradi have long whip-like structures whose function is unknown. In Uropygi the telson trails behind, but in Palpigradi it is arched and carried over the abdomen, like the metasoma of a scorpion.

In striking contrast to this there is no trace whatever of either a three-somite tail or a post-anal telson of any sort in Solifugae, Opiliones or Pseudoscorpiones. This shows that the latter structure has evidently been developed independently by the groups which possess it. Its origin may be sought in the pointed termination to the body or certain of Eurypterida, but its existence does not necessarily point to a closer relationship between the telson-bearing orders.

Thus there is a very considerable diversity in the form of the opisthosoma in the different orders of Arachnida, a diversity which is most puzzling in the foremost somites. If it is to be assumed that in all orders the genitalia open on the same somite, the eighth, then it follows that in Solifugae, Opiliones and Ricinulei there are not as many as 12 opisthosomatic somites in all, in the scorpions there are 11 post-genital somites and in the false scorpions, spiders and some others there are ten post-genital somites.

A general summary of the segmentation of the arachnid body is given in tabular form on p. 20.

In all ordinary Arachnida there are six pairs of prosomatic appendages.

The chelicerae, the first or pre-oral pair, consist of two or three segments. In the latter case the third segment meets a prolongation of the second, forming a chelate organ, useful for picking things up and perhaps biting their heads off. The two-segment chelicerae cannot nip but can only pierce. By their position in the front of the body the chelicerae are well placed to meet a variety of needs and perform a variety of functions, and indeed the versatility of the arachnid chelicera is equalled by but few arthropodan appendages. They are enormously developed in Solifugae, where they also carry out the vital operation of placing the packet of sperm in the vagina of the female, they are used by trap-door spiders for digging and by scorpions for making a noise. In false scorpions they contain silk glands and in spiders, poison.

SEGMENTATION OF THE ARACHNID BODY

Somite	
1.	Chelicerae
2.	Pedipalpi
3.	Legs 1
4.	Legs 2
5.	Legs 3 (Mesopeltidium)
6.	Legs 4 (Metapeltidium)
7.	Pre-genital or Pedicel
8.	Genital: Book-lungs 1
9.	Book-lungs 2: Pectines
10.	Book-lungs 3: Spinnerets 1
11.	Book-lungs 4: Spinnerets 2
12.	Tergite 6
13.	Tergite 7
14.	Tergite 8
15.	Tergite 9
16.	Tergite 10 ⎫
17.	Tergite 11 ⎬ Pygidium or post-abdomen
18.	Tergite 12 ⎭
Telson	Sting; Flagellum

The pedipalpi are appendages of six segments, the first of which, the coxae, most frequently have extensions, called maxillae or gnatho-bases, which function as mouth parts with or without contribution from the coxae of the anterior legs. The limbs themselves may be simple tactile organs outwardly resembling the legs, as in spiders, or chelate weapons of great size, as in scorpions and false scorpions. They may be specialized in different ways, as in spiders where they act as accessory male organs, and in Solifugae, where they terminate in suckers. It is to be noticed that the conspicuous and actively functional limbs are sometimes the chelicerae and sometimes the pedipalpi, but that both are not found enlarged in the same order.

The legs are of seven segments: coxa, trochanter, femur, patella, tibia, metatarsus and tarsus. It is the metatarsus that is missing from the pedipalpi. They may be all alike (and this is the general rule), but in some orders, notably Amblypygi and Solifugae, the first pair are not used for walking but are carried aloft and directed forwards as tactile organs.

Some confusion exists in the names applied to the segments of the legs in Scorpiones, Amblypygi, Uropygi and Palpigradi. The problem is a dual one. In some instances it is merely the use of a different word to describe the same thing, as for example the alternative of praetarsus or transtarsus to describe the extreme terminal portion of a limb.

In other instances there is a divergence of opinion as to the category to which a segment really belongs. For example, the fourth segment of the scorpion's leg is named the "tibia" by Kraepelin and the "patella" by Petrunkevitch. This is a more serious problem. A solution could be reached only by a rather lengthy discussion of the exact meaning of each of the words applied to leg segments, and would involve a discussion of the internal musculature and also a comparison with other classes of Arthropoda. This extremely interesting piece of work is, it must be admitted, disproportionate to the design and scope of this book. The solution seems to be that the most useful purpose will be served if the existing nomenclature be tabulated so that divergences are made evident. This will reduce the confusion that otherwise is likely to arise (see p. 22).

Legs also differ in other ways among themselves. The number of claws is not always the same on all, and there are differences in the number that bear gnathobases on the coxae. In Limulus and some of the Opiliones all the coxae assist in mastication; in scorpions only the first two pairs; and in spiders this duty is confined to the pedipalpi and the legs take no share in the work.

Like the other appendages, legs have several functions, even when climbing, seizing prey, digging and swimming have been added to their fundamental purpose of walking. They are richly provided with sense organs in the form of spines of different types, mentioned below, as well as tarsal organs which are chemoreceptors, and lyriform organs of uncertain function. It is thus no surprise that in some forelegs the sensory information-value has superseded the value as organs of transport, and by members of some orders they are carried clear of the ground. Of Opiliones it has been said that "the study of harvestmen is the study of legs", and of the spider that "while it is sufficiently true to the traditions of the animal kingdom to see with its eyes and taste with some part of its mouth, it hears and it feels and it smells with its legs".

In Arachnida the legs almost always, and the pedipalpi often, end in claws: curved, sharply pointed pieces of chitin which may be smooth or toothed. There may be one, two or three claws. The number of teeth varies over a wide range. Some of the claws on the legs of the spider Liphistius, for example, have one or two teeth, while the legs of the spider Philaeus may have 29. The number is not constant throughout

	1	2	3	4	5	6	7	
SCORPIONES	Coxa	Trochanter	Femur	Tibia	Tarsus i	Tarsus ii	Tarsus iii	Kraepelin
	Coxa	Trochanter	Femur	Tibia	Metatarsus	Protarsus	Tarsus	Birula
	Coxa	Trochanter	Femur	Patella	Tibia	Metatarsus	Tarsus	Petrunkevitch
"PEDIPALPI" Legs 2–4	Coxa	Trochanter	Femur	Patella	Tibia	Basitarsus	*Tarsus*	Börner
	Praecoxa	Transcoxa	Femur	Patella	Tibia	Co-tibia	*Tarsus*	Hansen
UROPYGI Leg 1	Coxa	Trochanter	Femur	*Patella-tibia*		Basitarsus	*Tarsus*	Börner
	Praecoxa	Transcoxa	Femur	Patella	Tibia	*Co-tibia*	*Tarsus*	Hansen
AMBLYPYGI Leg 1	Coxa	Trochanter	Femur	Patella	*Tibia*	*Basitarsus Tarsus*		Börner
	Praecoxa	Transcoxa	Femur	Patella	*Tibia*	*Tarsus*		Hansen
PALPIGRADI Legs 2–4	Coxa	Trochanter	Femur	Patella	Tibia	Basitarsus	Tarsus	Börner
	Praecoxa	Praefemur	Femur	Patella	Tibia	Co-tibia	Tarsus	Hansen
PALPIGRADI Leg 1	Coxa	Trochanter	Femur	Patella	Tibia	*Basitarsus*	*Tarsus*	Börner
	Praecoxa	Praefemur	Femur	Patella	Tibia	*Co-tibia*	Tarsus	Hansen

Segments in italics consist of two or more portions.

the life of the individual: it increases moult by moult in web spinners and decreases in the same way among hunters.

The appendages of the opisthosoma, though they may make a transient appearance during development, seldom exist in the adult. The exceptions are found in scorpions and spiders. In the former the somite immediately behind the genital carries a pair of highly characteristic organs, known as pectines (see Fig. 31) and found in no other order. Among spiders the appendages of the third and fourth abdominal somites are present as spinnerets, carrying the orifices of the ducts from the silk glands. Primitively there are eight spinnerets, but the number may be reduced to six or four.

In an important study of these glands, Marples (1967) has pointed out that the traditional view of the spinnerets as the exopodites and endopodites of the fourth and fifth somites cannot be maintained. There are too many differences both in structure and development between the spinnerets and the biramous limbs of the (not closely related) Crustacea.

He has also described a group of epiandrous glands, found in most male spiders. These add to the sperm web a small white mat, on which the drop of sperm is deposited.

The structure of the exoskeleton, as revealed by a transverse section is shown in Fig. 1. The value of a hard exoskeleton as a protection against injury and as support for muscles is obvious; the tissue also performs an important function in helping to conserve internal moisture and prevent desiccation. All terrestrial Arthropoda need to solve this problem, for life on the land encourages evaporation, and too great a loss of water, and the concentration of the body fluids which results, are quickly fatal. Arachnida have reduced the risk by adopting special

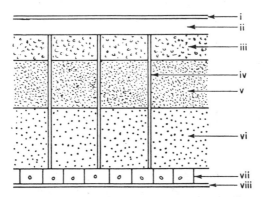

Fig. 1. Diagrammatic section of exoskeleton. (i) Cement layer; (ii) wax; (iii) cuticulin; (iv) pore canal; (v) exocuticle; (vi) endocuticle; (vii) hypodermal cellular layer; (viii) basement membrane; (i) + (ii) + (iii) = epicuticle.

modes of life and by the existence of a layer of wax in the epicuticle.
The result is a considerable tolerance of unfavourable conditions, but a
far from perfect adaptation, so that to be too dry is worse for an
arachnid than to be too wet.

The exoskeleton is also the site of sense organs, such as the eyes,
which are of limited use, and setae, which are of the greatest value as
organs of touch, as well as special chemotactic organs and others which,
like the lyriform organs, are mysterious and puzzling to us.

The eyes or ocelli are nearly always of the smooth or simple type, with
a lens which is a curved and transparent portion of the cuticle. Their
number varies from two to 12, and they are usually sessile, or level
with the surface. Stalked eyes are unknown, but in several families,
especially among the spiders and harvestmen, they are raised on a
turret or ocular tubercle. The general structure of an arachnid eye is
shown in Fig. 2.

The eye consists of the following parts: (i) the lens, described above,
a curved portion of exoskeleton free from setae and pigment; (ii) the
vitreous body, lying beneath the lens; (iii) a pre-retinal membrane, a
prolongation of the basal membrane of the hypodermis; (iv) a retina of
visual cells, processes from which make contact with the optic nerve;
(v) a post-retinal membrane, sometimes pigmented, forms the back of
the eye. There may also be pigment cells or rhabdomes among the

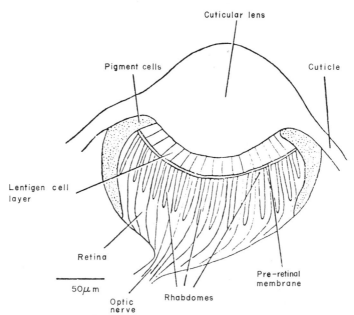

FIG. 2. Vertical section through the eye of a harvestman. After Curtis.

cells of the retina and a tapetum or reflecting layer below them.

Considerable variation of these parts, and especially in the shape of the lens and the extent of the vitreous body, are to be found in different orders and families, but the chief variation is in the position of the visual cells of the retina. These cells may be likened to a pear with a long stalk. In post-bacillar eyes the swollen base, which contains the nucleus, lies on the post-retinal membrane with the stalk directed outwards towards the vitreous body; in pre-bacillar eyes the pear is the other way round and the nucleus is nearer to the lens than the stalk. These are the eyes which normally have a tapetum, reflecting the incident light so that it passes through the visual cells a second time.

Both kinds of eyes are found in most orders of Arachnida. Their value as organs of distinct vision is limited; only jumping-spiders, and to a lesser extent wolf spiders, can form anything like a clear image. The eyes have different outward appearances, which have caused them to be given a number of contrasting names, such as diurnal and nocturnal eyes, direct and indirect, principal and secondary, median and lateral and so on.

On the surface of the body and limbs of Arachnida a number of important structures are visible and may be described as follows.

Tubercles

These are hollow outgrowths of the cuticle, such as occur on the hoods of Trogulidae and on the pedipalpi of many other harvestmen. They are not sense organs. Similar outgrowths, but solid throughout, are often known as denticulae, and when very short may be distinguished as spicules.

The remaining structures are sensory in function.

Spines

These are solid bristles, usually black in colour and very obvious on the legs of nearly all Arachnida. They may arise directly from the cuticle, but they often originate from the top of a hollow tubercle. The arrangement of these spines is constant for a species and is often used in taxonomy. Under high magnification it can be seen that the surface is sculptured with a spiral marking, which sometimes produces the appearance of a saw-edge. The chitin below is in layers and often the centre is palest, so that a spine seems to be hollow. Spines arise from a colourless area of the exoskeleton, surrounded by a darker ring and the shape of their bases is always a characteristic (Fig. 3). Each spine is in communication with a nerve ending, and it has long been known that they are erectile. In male spiders they can be seen moving during mating.

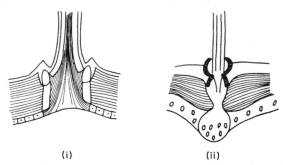

(i)   (ii)

FIG. 3. Setae: (i) from Biton (Solifuge) after Bernard; (ii) from Anomalobuthus (Scorpion) after Pawlowsky.

Setae

Setae (chaetae) are finer than spines and often lie closer to the surface instead of standing at right angles to it, and unlike spines, which form rows, setae grow in patches, covering an area. The overall hairiness in the appearance of an arachnid leg is due to its clothing of setae, which may be plumose. Setae are not erectile, their bases lie inside rings surrounding pores in the cuticle. Microsetae resemble them but are smaller.

Trichobothria

These are the finest of all. They are more or less erect, and are much longer than setae, among which they sometimes grow. They are sensory receptors, and are as characteristic of certain orders as are the spider's web and the scorpion's sting. They arise from two or three cupules in the hypodermis (Fig. 4).

First described and named Horhaare by Dahl (1911) they have not attracted intense research until recently, it being generally believed that they were stimulated by the slightest movements of air, including infrasonic and ultrasonic sound waves. This supported the contemporary belief that spiders appreciate music; an idea that suffered from the observations of McCook. He had detected a reaction to the notes of a flute, and later found the same response to a silent puff of air.

An important function follows from their response to air movement such as is produced by the passing of other animals nearby. A false scorpion can, by their means, detect the presence of prey at a distance of 15 mm. A scorpion such as *Centruioides sculpturatus* sits motionless in the dark, prepared by the same stimulus from its trichobothria to pounce upon an approaching insect. Each trichobothrium has been shown to oscillate in one plane only, so that a set of them collectively indicate the direction from which air disturbance is coming. Thus they have an

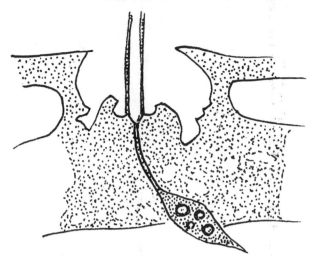

FIG. 4. Trichobothrium. After Gossel.

orientation function. A further suggestion is that they react to the light of the sun and the moon, and so are described as astrotactic. This is in addition to their sonotactic function, now well established.

These remarkable organs occur on the legs and pedipalpi of spiders, on the pedipalpi of false scorpions and scorpions, on the tarsi and patellae of the three orders of "Pedipalpi", and on the bodies of certain mites. They are not found on Solifugae, Opiliones or Ricinulei.

Gabbutt (1972) has shown that the trichobothria of pseudoscorpions form a valuable guide to the genera and species in this order; and further that they help to distinguish between the nymphal stages of immature specimens. Most recently the trichobothria of scorpions have been exhaustively studied by Vachon (1973). Among his conclusions is the important fact that the numbers and positions of all trichobothria are established among the scorpions at the early stage of the first nymph, and that this is a distinction from the state among pseudoscorpions. Each trichobothrium is to be regarded as an individual sense organ, to be given and described by its own symbol. Thus there develops a trichobothrial nomenclature. It is clear that trichobothriotaxy has great taxonomic and phylogenetic significance.

The exoskeleton of an arachnid also carries other sense organs of types that are less familiar because they have no analogues among the vertebrates. The most plentiful and the most mysterious of these are the lyriform or slit sense organs, the outward appearance of which is shown in Fig. 5. Under a microscope they look like slits in the cuticle.

They may be very numerous, for, either as single slits or as compound

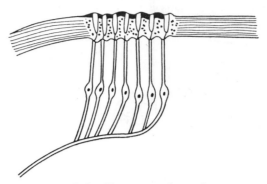

FIG. 5. Lyriform organ in section.

"slit sense organs" they occur on the carapace and on the sternum and on the pedipalpi and legs, especially near the joints. Barth and Seyfarth (1972) estimate as many as 3,000 on the surface of a single spider.

During a long period of uncertainty several functions have been attributed to these objects. They have been described as auditory organs, chemoreceptors, detectors of heat and of atmospheric moisture. Pringle (1955), developing the idea that they are similar to the campaniform organs of insects, described them as modifications of the cuticle near the limb joints, with nerves or "sense cell processes" attached to the centre of a thin membrane. Thus he introduced the conception of mechanoreceptors, affecting orientation. Working with amputated legs from scorpions and Amblypygi, he attached to the bases of the nerves platinum electrodes connected to an amplifier and an oscilloscope. He found that pressure on the limb produced large impulses, and concluded that the lyriform organs supply a kinaesthetic sense, comparable to the muscle-tendon senses of vertebrates. Such a sense would play an important part in the phenomenon of Netzstarr-heitstaxis of Holzapfel (1933).

Similar results were obtained by Edgar (1963), using the legs of six species of Opiliones.

Barth and Seyfarth (1972) have obtained striking results in favour of the orientation theory. In their apparatus a live spider, *Cupiennius salei*, was attracted by the buzzing of a bluebottle at distances of up to 30 cm. On reaching the fly the spider was chased away and the fly removed. In normal circumstances and even with its eyes covered, the spider soon returned to the site of the insect. By immobilizing the lyriform organs on the femur or tibia the authors proved that these organs controlled kinetic orientation, a term that implies the determinaton of an animal's direction by its previous movements. Continuation of this work at Munich has shown that the organs are strictly mechano-receptors, and that the minute deformations of the membrane, on which they depend, may be produced by:

(i) vibration of the substrate;
(ii) air-borne sounds;
(iii) movements of parts of the body;
(iv) the motion of walking.

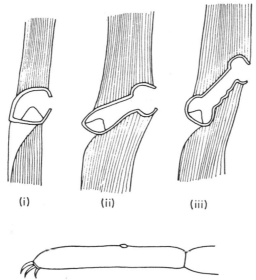

FIG. 6. Tarsal organs of spiders: (i) Myro; (ii) Bomis; (iii) Synaema.

The tarsal organs (Fig. 6) are chemoreceptors and their use has been satisfactorily demonstrated by Blumenthal (1935). These organs can be seen under the microscope as small round holes in the upper surface of the tarsus, leading into a depression at the bottom of which lies a small projection. They are used in the testing of drinking water and the examination of edible prey. If the tarsus of one of the anterior legs of a thirsty spider is touched with a drop of water, the spider moves forward and drinks; if one of the posterior legs be so touched, the spider turns round and drinks. If the tarsal organ is sealed these responses disappear, and if a drop of water touches any other joint of the leg, the spider merely moves away. Blumenthal also showed that the organ enabled a spider to distinguish between water, brine, sugar solution and quinine.

There is abundant evidence that most Arachnida react to sounds, yet arachnologists have always been loath to admit that spiders can hear anything or that scorpions and Solifugae produce during stridulation sounds that are audible to other members of their species. Part of this reluctance depends on the interpretation of the word "hear", and part on the strange mechanism of the sense organs involved. In the parallel case of tasting or smelling an acceptable description of the organs has been found: they are called chemotactic. Similarly, the setae or the trichobothria, by which Arachnida almost certainly are caused to respond to vibratory stimuli, might with equal acceptability be called "sonotactic".

4

Physiology: Internal Organs

As is only to be expected of an animal as highly specialized as an arachnid, the internal structure of the body is complicated. It is best described by the conventional method of considering it as made up of a number of organ systems, each with its own functions; but it is important for the reader to remember that in the living animal the functioning of each of these systems is dependent on the rest. The systems are:

The alimentary system	The nervous system
The respiratory system	The reproductive system
The vascular system	The excretory system
The glandular systems	The muscular system

The alimentary system follows the pattern common to all Arthropoda in that it consists of fore-gut, mid-gut and hind-gut, the first and the last of which are lined with a chitinous invagination of the exoskeleton. The mouth, whose characteristic position behind the chelicerae has already been mentioned, usually lies above and between the coxae of the pedipalpi. It is bordered above by an upper lip, the epistome or rostrum, and below by the lower lip or labium, derived from a sternite (Fig. 7). Either or both of these, as well as the maxillary lobes of the pedipalpi, may contain glands, the secretion from which is poured into the prey. In Scorpiones, Pseudoscorpiones and Solifugae the rostrum and the pedipalpal coxae form a space or atrium in front of the mouth. The floor of this atrium is the labium, curved upwards and set with setae, the whole forming a filter through which fluid nutriment only can make its way. In Uropygi, Schizomida and Ricinulei the pedipalpal coxae meet and fuse in the middle line, forming a plate which is concave on its dorsal surface. Into this fits the convex rostrum. Both surfaces are set with spines or spikes, which form a filtering device. The mouth opens into this space, which is known as the camarostome. A similar fusion of the palpal coxae occurs in the extinct Kustarachnae.

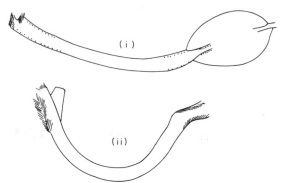

FIG. 7. Fore-gut of a spider. (i) Oesophagus and stomach of Araneus; (ii) oesophagus of Tegenaria.

Filtering of the imbibed nutrients occurs in spiders in the pharynx; in scorpions in the channel between the long maxillary processes of the first two pairs of legs; in the Opiliones there is no filtering.

The pedipalpal coxae, which contain the maxillary glands in spiders, play no part in feeding in Scorpiones, Pseudoscorpiones or Solifugae. In general, the prey, held in the chelicerae, is drenched by the salivary juices from the glands just mentioned, and the solution resulting from this external digestion is sucked in by the action of the expansible pharynx. Near the opening of the pharynx a patch of sensory cells enables the animal to detect unpleasant tasting fluids and then eject them if necessary.

The pharynx is followed by a narrow oesophagus which passes through the nerve collar and, in spiders and some other orders, ends in an expanded bulb, the pumping pharynx, sometimes inaptly called the sucking stomach. This is attached by muscles to the carapace above and the endosternite below, and the contractions of these muscles produce a rhythmical sucking action, sometimes to be seen when an arachnid is feeding. This organ marks the end of the stomodaeum.

The mid-gut or mesenteron is soft-walled and usually extends past the middle of the opisthosoma. Its very remarkable characteristic is the large number of branching diverticula that arise from it. These diverticula have been fully studied in spiders by Millot (1931), who found that they may be divided into four types. In the first or simple type there are two short sacs directed forward (Fig. 8). Each has a more or less clearly defined secondary portion, but the two sides are not always symmetrical. The second or intermediate type shows three or four diverticula on each side, by no means clearly divisible from one another, and not reaching the coxae of the legs. In the third or classical type (so called because the earlier workers described this type only) there are

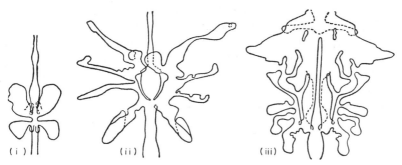

FIG. 8. Prosomatic diverticula of alimentary canal (from Millot): (i) Scytodes; (ii) Zoropsis; (iii) Ballus.

five pairs of diverticula, one anterior and four lateral, reaching the bases of the legs and entering the coxae for a short distance (Fig. 8). The customary statement that the anterior diverticula meet and form a complete ring is incorrect. The fourth or complex type differs from the third in a great development of either the anterior or the lateral diverticula, so that these divide or branch, and are so large that they quite change the appearance of the interior of the prosoma (Fig. 8). Millot found that although individual modifications are common, the type of arrangement is constant within the limits of a family.

Mesenteron diverticula are very numerous in the opisthosoma, where their branches surround the other organs and form most of the bulk of the hinder parts of the body. Petrunkevitch calls this region of the system the chylenteron.

The partially digested mixture of food and saliva that enters the mesenteron soon passes into these diverticula, where two important types of cells are found in the epithelium: the secretory cells and the absorptive cells. From the former there comes an albuminous mixture which, continuing the digestive process, produces solutions which pass into the adjacent absorptive cells. Here the final digestion occurs; fats and albuminoids, which can be assimilated or stored, are formed and removed by the blood, while urates and other excreta remain in the cell to be ultimately eliminated.

The many branches of the mid-gut diverticula thus act in part as a great storage organ, and to this extent justify the old name of "liver".

They endow Arachnida with their remarkable ability to fast for long periods, living on the stores they preserve.

The last part of the mesenteron receives the Malpighian tubules of the excretory system.

The hind-gut or proctodaeum is, in general, a short tube leading to the anus, but in some pseudoscorpions it is unexpectedly long and curved. It often includes a dilated portion, the stercoral pocket, in which faecal matter and excreta temporarily accumulate.

Among Arachnida there are two types of respiratory organs or oxidizing devices, known as book-lungs (or lung-books) and tracheae. There are therefore four kinds of respiratory equipment, constituted of either or both or neither of these devices.

Tracheae alone are found in Pseudoscorpiones, Solifugae, Opiliones, Ricinulei and Acari: the scorpions, Uropygi and Amblypygi have book-lungs only; a few families of spiders have tracheae only, but most spiders possess organs of both kinds; the small Palpigradi, the Cyphophthalmi and some of the Acari have only cutaneous respiration.

The two systems received exhaustive study from Levi (1967). He pointed out that the tracheae are of two kinds, sieve tracheae and tube tracheae. The former exist in bundles that arise from a tubular origin on the eighth somite, the position alternatively occupied by the anterior pair of book-lungs. They are found in pseudoscorpions, Ricinulei and in some spiders. Tube tracheae are usually unbranched and are present in Opiliones, Solifugae and in most spiders. It is evident that book-lungs and tracheae are homologous, both being formed by ectodermal invagination. Wherever a tracheal system is elaborate it is probably more efficient than book-lungs, as is shown by their sole presence in Solifugae, the most active of all Arachnida, and in the active Opiliones, which may also have tracheal openings on the legs.

When book-lungs are formed by invagination, they develop a number of folds on the anterior surface, which project into the lung cavity and thus roughly resemble the leaves of a book (Fig. 9). There are one, two or four pairs of lungs, always situated in the anterior part of the opisthosoma. There are no respiratory movements; air enters the pulmonary spiracles and reaches the blood through the delicate epithelium of the leaves. The number of leaves ranges from about five to about 150. Species with many tracheae have fewer leaves.

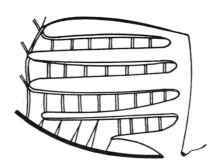

FIG. 9. Diagram of a book-lung, showing only four leaves.

The following notes may be added to the above statements. Scorpions all have the full provision of eight book-lungs, yet they have so small a need for oxygen that seven of these can be plugged or sealed without apparent effect. In spiders the relative development of lungs and tracheae is very different in the different families, and does not seem to be related to the mode of life. The simple idea that great activity demands a copious supply of oxygen does not seem to be supported. Yet in the Solifugae there is a greater elaboration of the tracheae than in any other arachnid. There are three pairs of spiracles, one of which is in the prosoma, and in one family there is an additional median spiracle in the posterior part of the opisthosoma. The tracheae themselves are wider, proportionately, than those of other Arachnida and they branch copiously throughout the entire body, appearing almost to have a segmental pattern. The spiracles are provided with muscles by which they can be opened or closed. In all, there seems to be no reason for not associating this unusual system with the greater speed with which the Solifugae are at times inclined to move.

The blood or haemolymph of an arachnid is an almost colourless fluid in which corpuscles of three kinds are to be seen. There are small clear leucocytes of a basophile character, some larger hyaline phagocytes and also large acidophile granular leucocytes. In spiders and scorpions the fluid itself is poisonous and can kill small mammals when it is injected into them.

The heart (Fig. 10) is tubular and lies in a pericardium in the mid-dorsal line of the opisthosoma. In Uropygi and Solifugae a part of it extends forwards into the cephalothorax. The blood runs forwards in a single anterior artery and backwards in three posterior arteries, and after passing through the smaller vessels, reaches the book-lungs. Aerated, it collects in a ventral sinus and so re-enters the pericardium through ostia that are closed by valves. The number of the cardiac ostia varies from one to nine pairs. Contracting ligaments effect the pulsating of the heart, which in scorpions may beat from 60 to 150 times a minute. The heart of a scorpion is larger and the vascular system more fully developed than in the other orders; it is much reduced in the Solifugae, in correspondence with the great dependence on tracheal respiration. In Opiliones it is short and has only two pairs of ostia. Among spiders the development of the heart and the number of pericardial ostia is more variable than in any other order, and Petrunkevitch has used this character as an important feature in the classification of the group.

The nervous system in Arachnida follows a pattern familiar in other classes of Arthropoda. It consists of a cerebral mass united by circum-oesophageal commissures to a ventral nerve cord, which primitively

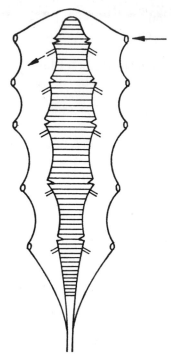

Fig. 10. Diagram of heart in pericardium, with five ostia.

and in most embryos swells to a ganglion in each somite. Considerable evolutionary changes in this plan have taken place.

The cerebral mass is usually regarded as having been derived from the protocerebrum and tritocerebrum found in polychaete Annelida, the deutocerebrum having disappeared. This loss may be associated with the absence of antennae and the gradual lessening of functional importance of the eyes. The chelicerae are innervated from the tritocerebrum.

The sub-oesophageal mass represents a number of fused ganglia which move forward during development. The number involved varies according to the order, and, as might be expected, the closest approach to the primitive condition is found in the scorpions. Here the mass is composed of nine ganglia, and the ventral nerve cord which follows has three ganglia in the prosoma and four in the opisthosoma, the last being a double one. When the ganglion which innervates the chelicerae and which forms a part of the cerebral mass is included, this gives a total of 18 ganglia in all. The same total is reached in four other orders:

	CHELICERAL	SUB-OESOPHAGEAL	OPISTHOSOMATIC
Schizomida	1	9	8
Uropygi	1	12	5
Amblypygi	1	17	0
Araneae (Liphistiidae)	1	17	0

In the remaining spiders and in the other orders the evolution of the system has continued (Fig. 11), and in Pseudoscorpiones, Acari and Opiliones the sub-oesophageal ganglion represents all the posterior ganglia in one: the opisthosoma has no central nerve cord and no ganglia.

Neurosecretory cells, producing hormones, have been found in all Arachnida that have been examined with sufficient precision.

The reproductive organs of Arachnida present an unusual mixture of possible variations, for the general plan of their parts is characteristic of almost every different order. An outline of these variations is shown below:

	FEMALE	MALE
Scorpiones	A network of tubes, consisting of three or four longitudinal ducts, united by transversals	A similar set of ducts, the tubes being rather narrower
Pseudoscorpiones	A median ovary, covered with follicles, united to a pair of oviducts	One median ventral testis with two vasa deferentia
Solifugae	Paired ovaries with follicles on the outer sides and a pair of oviducts	Four longitudinal testes, their ducts uniting to two and then one
Amblypygi	A median ovary and two oviducts	Two broad tubular testes
Araneae	Two ovaries, conspicuously covered with follicles, two oviducts uniting to a median vagina (Fig. 12)	Two testes and a pair of vasa deferentia
Opiliones	A horseshoe-shaped ovary, its anterior ends joined to a median oviduct	A U-tube shaped testis, with median vas deferens to penis

These and other differences are to some extent related to the varied methods of insemination, which are themselves part of adaptation to life on the land. In some orders, such as the Pseudoscorpiones, the male secretes material which hardens to form a pillar-like spermatophore, on the top of which a packet of spermatozoa is placed. The female is induced to take these gametes into her vagina.

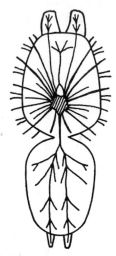

FIG. 11. Nervous system of a spider.

This is regarded as the more primitive method. In some orders it has been replaced by adapting a limb for conveying the sperm: the pedipalpi in Araneae, the third leg in Ricinulei. Lastly, there may be a real intromittent organ and copulation in the true sense may occur, as in the Opiliones and some Acari.

Three distinct methods of excretion are found among Arachnida, the most fully developed of which is the system of coxal glands (Figs 13 and 14). In their typical form, many modifications of which are, however, found, they consist of a large excretory saccule, lined with

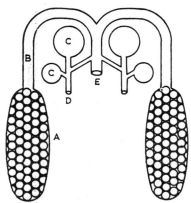

FIG. 12. Diagram of female reproductive organs of a spider. A, ovary; B, oviduct; C, spermatheca; D, orifice for entry of sperm; E, vagina.

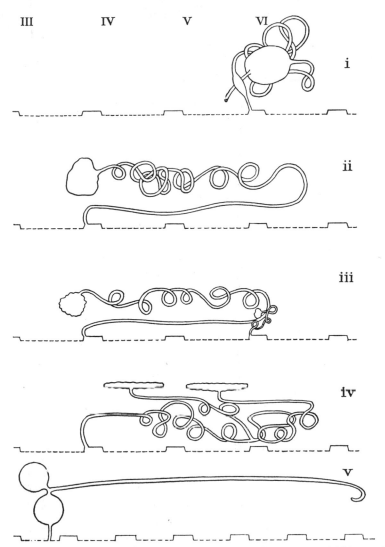

III   IV   V   VI

i

ii

iii

iv

v

FIG. 13. Coxal glands. (i) Scorpiones; (ii) Amblypygi; (iii) Amblypygi (Charontini);
(iv) Uropygi; (v) Palpigradi.

cubical or flattened epithelium, lying outside the endosternite opposite
the coxae of the first legs. This discharges its products into a convoluted
tube, the labyrinth, the coils of which occupy the space from the first to
the fourth coxa, or beyond. At the distal end of the labyrinth there is
sometimes a swelling, regarded as a "bladder", and from this there runs
forward a straight tube, the internal limb of the labyrinth, which lies

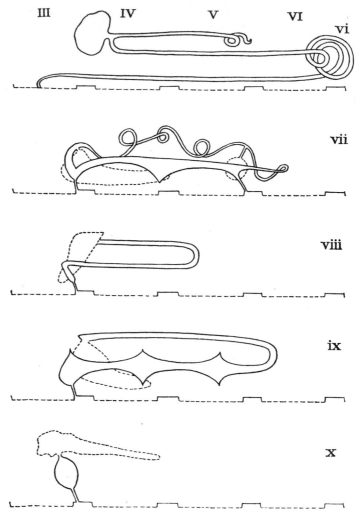

FIG. 14. Coxal glands. (vi) Solifugae; (vii) Theraphosomorphae; (viii) Dysderidae; (ix) Lycosidae, Agelenidae, Thomisidae; (x) Araneidae, Pholcidae.

inside the convoluted portion and from which short exit tubes open to the exterior at small orifices behind the first and third coxae.

In Solifugae and Palpigradi, there is an additional tube, the labyrinth sac, lined with secretory cells, between the saccule and the labyrinth. The orifice in these forms is on the pedipalpal somite.

The following table shows the variations which occur among the different orders of Arachnida. The most variable order is the Araneae,

	SACCULE IN SOMITE	LABYRINTH SAC	LABYRINTH	EXIT
Scorpiones	5 and 6	Absent	Coils in somite 5	Legs 3
Amblypygi	3	,,	Extensive coils to somite 6	Legs 1
Uropygi	4 and 5	,,		Legs 1
Araneae Theraphosomorphae	3 and 5	,,	Large, coiled	Legs 1 and 3
Araneae Gnaphosomorphae	3	,,	Straight tube	Legs 1
Palpigradi	2	Extending to somite 8	Small vesicle	Palpi
Solifugae	2	Extending to somite 4	Coils to somite 6	Palpi

in which four types exist showing a progressive simplification, correlated with a corresponding increase in complexity of the silk glands.

The Malpighian tubes are of interest because they are a parallel to the similar tubes of Insecta but are not their homologues, for the tubes in Arachnida have a hypodermal origin, compared with the ectodermal origin in Insecta. They are the chief excretory organs of Arachnida: they branch copiously among the many mid-gut diverticula from one or two points of origin on the posterior end of the mesenteron. Their epithelial cells absorb waste matter from the haemocoele, transform it and excrete it in the form of guanin to the hind-gut. Here it mixes with the faecal residues in the stercoral pocket until ejected, either periodically or, sometimes, at moments of shock or stress.

Malpighian tubes do not generally enter the prosoma. In this region cells known as nephrocytes absorb the products of metabolism from the blood sinuses. Other types of cell, the hypodermic cells below the exoskeleton and the superficial and interstitial cells round and between the intestinal diverticula also play a part in the absorption of excreta.

The bodies of Arachnida are very fully provided with glands. There are venom glands in Scorpiones and Pseudoscorpiones, silk glands in Acari, Pseudoscorpiones and Araneae, acid glands in Uropygi and odoriferous glands in Opiliones. The glandular systems involved are usually peculiar to one order, so that fuller accounts of them will be found in the appropriate chapters in Part III.

Since nearly all the movements of an arachnid are movements of its legs and its mouth parts, it is not surprising that its muscles are very unequally divided between the prosoma and the opisthosoma. Within the former there is a conspicuous endosternite, a plate of chitin to which most of the muscles are attached. Typically four pairs of bands of muscle fibres run upward from the endosternite to the carapace, and, in spiders, a median band in addition. A varying number of compensating muscles join the sternum to the endosternite, round which a series of lateral muscles run horizontally.

This full set of muscles effect the movements of the appendages, the gut and the prosomatic glands.

In the opisthosoma the most significant muscles are those which, in narrow strips, run vertically between the dorsal and ventral surfaces. They are manifestly metameric, and must be presumed to be vestiges of muscles which formerly united all tergites and sternites. The posterior pairs have vanished from modern Arachnida, and the number remaining is closely related to the visible segmentation of the opisthosome. Thus there are eight pairs in Amblypygi, seven in Schizomida and six in Solifugae and Palpigradi. Other muscles in the opisthosoma are concerned with the moving of the alimentary canal and the opening and shutting of the different orifices.

The muscles of a spider's leg, which may be taken as an example of

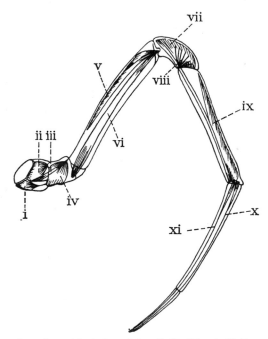

Fig. 15. The muscles of a spider's leg. After F. D. Wood. (i) Extensor trochanteris; (ii) flexor trochanteris; (iii) flexor longus femoris; (iv) flexor bilobatus femoris; (v) flexor bilobatus patellae; (vi) flexor patellae; (vii) protractor tibiae; (viii) flexor tibiae; (ix) flexor metatarsi; (x) extensor tarsi; (xi) flexor tarsi.

an arachnid appendage, are shown in Fig. 15. Their most interesting feature is the extreme development of a delicate muscular sense, shown so dramatically by the habits of web spiders. Their responses to the varying tensions of the threads of their webs are among the most amazing items in arachnid behaviour, and are responsible for the vibrotaxis and Netzstarrheitstaxis described later.

The work of Parry and Brown (1959) has shown that in jumping these muscles are not solely responsible for the spider's leap, but that much of the motive power is supplied in the form of hydraulic pressure from within. Thus there is no exaggeration of one pair of legs, such as is found in many other invertebrates that jump, and a unique method of using all pairs of legs simultaneously has taken its place.

5

Embryology: Development

In all Arachnida the male and female sexes are recognizable and ova normally develop only after fertilization by a spermatozoon. The course of gametogenesis has been followed in only some of the orders.

Oogenesis produces both ova and yolk, the former being produced after the casting out of two polar bodies from the secondary oocytes. Much of the cytoplasm in an ovum is converted into yolk, which appears first in the form of droplets. These collect in lines radiating from the egg-nucleus, which is drawing nourishment from the surrounding lymph.

Spermatogenesis often includes a proportion of amitotic division, resulting in the formation of two types of spermatozoa, both of which have been seen in the spermathecae of spiders. This amitosis was described by Warren in 1925, and a parallel formation of two kinds of spermatozoa has been demonstrated by Juberthie (1964) in Cyphophthalmi. The phenomenon is also known among insects and molluscs.

The diploid number is not the same in all orders, nor even in the same order. In the scorpion Buthus it is 20 to 22, in Opisthocanthus it is 80 to 100. Yaginuma (1964) recorded for three species of scorpion the haploid numbers of 27, 12 and 11. In a table summarizing results from other species this number varied from 3 to "about 60".

In both Uropygi and Amblypygi 24 autosomes have been counted, together with one X-chromosome. Among spiders, counts have given 18 for Anyphaena and 48 for Dugesiella. The chromosomes of 57 species of spiders from 17 families have been described by Suzuki (1954). He found that the haploid number varied from 4 to 24, except in Heptathele, where it was 48. He further noted that the number was higher among the more primitive families of the Theraphosomorphae than the more specialized families, where 15, 13 and 12 were by far the commonest counts. Generally the spermatocytes carry two X-chromosomes, X_1 and X_2, which are together present in half the spermatozoa, but in a few species three or even four X-chromosomes were detected.

In Opiliones the diploid number varies from 32 to 16, and there may be one X-chromosome.

For many years there have been reports of apparent parthenogenesis among Arachnida. Campbell (1884) wrote of a *Tegenaria parietina* which spent much of its life in captivity and can have had no chance of meeting a male. From a cocoon that she made two young spiders emerged. Damin (1894) told of a *Filistata testacea* which moulted twice in captivity and laid a number of eggs, of which 67 produced normal nymphs.

There have been several records of incomplete development of unfertilized ova and a most interesting account by Machado (1964) of the tiny spiders Theotima. No males and no spermatozoa in the spermathecae of three species of this genus have ever been seen, and after laboratory investigation Machado is confident that this is a case of genuine parthenogenesis.

Relative scarcity of males is well known in the harvestman *Megabunus diadema*. Phillipson (1959) took one male and 406 females of this species and kept ten of the latter under observation. Thirteen batches of eggs were produced and young nymphs hatched from them. Equally convincing reports of parthenogenesis or partial development of unfertilized eggs have referred to *Phalangium opilio*.

All reliable investigations of the proportions of the sexes in the recently hatched young agree that their numbers are approximately equal, and it may therefore be assumed that the sex of the individual is determined in the usual way by the chromosomes of the male gametes.

Fertilization occurs during the process of egg-laying, when the sperm are released from the spermathecae, enabling one of them to enter each egg before the hardening of the chorion and vitelline membrane.

The development of the zygote which now follows cannot be adequately described in general terms. The most detailed researches have all been devoted to the eggs of spiders, but at almost every stage the process is different in one of the other orders, and the embryology of some of these is at present but partially or scarcely known. A selection of topics may well give a better idea of the diversity of arachnid embryology than any attempt to construct a continuous narrative, bristling with exceptions.

The early divisions serve to underline this statement, since at least four kinds of segmentation have been described. Total segmentation is confined to a few mites and scorpions; among spiders and false scorpions segmentation is at first total, but is soon replaced by merely superficial divisions, and this superficial division is found in most of the other orders, as it is among insects. In some scorpions segmentation is discoidal.

As a rule the first division of the egg is meridional, producing anterior and posterior cells, a slight flattening of which indicates the future ventral surface. The second division is also meridional but at right angles to the first, and the third is equatorial. Thus an eight-celled stage consisting of four dorsal and four ventral cells is reached. The irregular dividing of the cells which now follows is accompanied by a movement of the cells between the yolk masses to the periphery, so that ultimately the embryo consists of perhaps 100 cells surrounding the yolk within. This stage may be called the periblastula.

The process of gastulation is represented by a multiplication of some of the cells of the ventral surface, where, at a point near the anterior end there appears an opaque, slightly projecting mass. This is called the anterior cumulus; its cells continue to divide, absorbing yolk as they do so, and ultimately come to form the mesoblast and hypoblast. In some orders a posterior cumulus arises at the same time and in the same way, in others the anterior cumulus divides into two. The subsequent meeting of the two cumuli or the division of the only one is the first distinction between the prosoma and opisthosoma of the future (Fig. 16A).

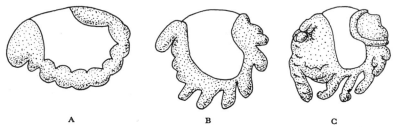

A   B   C

Fig. 16. Three stages in the embryonic development of a spider, Segestria. After Holm.

The appearance of two of the germinal layers in this way is followed by the formation of a temporary coelom, consisting of a series of metameric sacs. At this stage the division of the developing body into somites is clear (Fig. 16B).

The appendages begin to be formed at about this stage (Fig. 16C). The chelicerae appear first, to be followed by the rest in succession, from before backward. The chelicerae and pedipalpi are at first behind the mouth, the former assuming their pre-oral condition later. Appendages of the opisthosoma, though normally absent after hatching, make a temporary appearance in varying numbers as rudimentary knobs. They are most numerous in Solifugae, where nine or ten pairs of small tubercles are to be seen: they are fewest in Pseudoscorpiones and Opiliones (Fig. 17), which show only four pairs.

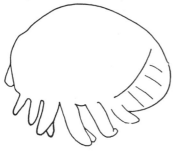 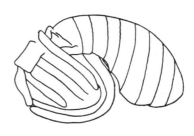

FIG. 17. Two stages in the embryonic development of a harvestman. After Holm.

The course of embryonic development in Solifugae, Uropygi and spiders, produces an organism which can be described as wrapped round the central mass of yolk with its dorsal surface innermost and concave and its ventral surface outermost and convex. This position has to be altered by a process known as inversion or reversion.

At or soon after the time at which the appendages make their appearance, the embryo, which has the form of a strip of developing cells, divides longitudinally into two similar halves, united only at the anterior and posterior ends. The two halves move apart, shortening the axis of the body, bringing its two ends closer together on the ventral side with a corresponding lengthening of the dorsal aspect. At the same time the two halves fold lengthways, enabling their edges to meet and unite, re-forming the dorsal surface. The whole process of reversion is a remarkable one, and gives an inevitable impression of a hasty attempt to repair a mistake that had occurred earlier in the development.

With the completion of the formation of at least the greater part of the systems of internal organs, the arachnid is ready for hatching. It is still surrounded by the inner vitelline membrane and the outer chorion, and these must be split. This splitting is nearly always facilitated by the use of an egg-tooth, a small hard projection either on the chelicerae or on the clypeus of the animal. Usually it carries an adequate quantity of yolk, which enables it to survive until it undergoes its first ecdysis and is, normally, able to feed itself; but most interesting exceptions to this are found in Scorpiones (Fig. 18) and Pseudoscorpiones.

The eggs of scorpions are yolkless, or nearly so, by the time that the embryo has completed development to the stage corresponding to reversion. The embryos develop each in a separate diverticulum of the tubular ovary, and the distal end of the diverticulum is drawn out into an appendix, which lies freely in the haemolymph and absorbs nourishment from it. The middle portion of the appendix serves as a reservoir, while the proximal end is a hollow chitinized teat, inserted into the

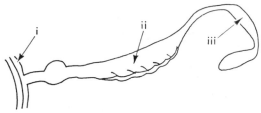

FIG. 18. Feeding mechanism of embryo scorpion. After Mathew. (i) ovarian tubule;
(ii) embryo; (iii) appendix.

mouth of the embryo. The embryonic chelicerae are modified to grasp
this teat between chitinized grooves in the specialized third somite.
Activity of the embryonic pharynx shows that the nourishment is
actually sucked in.

The development of a pseudoscorpion is characteristic. Two pre-
larval stadia are distinguishable. At an early stage a pumping organ is
formed in connection with the pharynx of the embryo. This is a
muscular sac which forces nutritive fluid into the embryonic gut during
the whole of the gestation period. At hatching it becomes converted
into the mouth parts of the protonymph, which is retained in the
brood-sac or incubation chamber.

This is a peculiar feature of the order. Just before the fertilized ova
travel down the oviduct the accessory glands secrete through the
genital orifice a membrane which forms an incubation chamber. The
eggs are laid into this, in numbers which vary between two and 50 or
more. The incubation chamber remains in position below the opisthosoma and is still in connection with the ovary, of which Vachon aptly
describes it as an external diverticulum.

Within this chamber or brood-sac the embryonic development of the
zygotes takes place. When the young are so complete as to be recognizable as false scorpions, they are still attached to the genital orifice of
the female, and are fed by the forceful actions of their mother on a
product of the ovary called "false scorpion milk". The little animals
swell as a result to three times their former volume and the incubation
chamber projects bulkily on each side of the mother's opisthosoma.
Ecdysis follows, producing the protonymph, which leaves its mother
and makes its way out of the breeding cocoon to begin life on its own.

6
Ontogeny: Growth

The hatching of an arachnid from its egg consists of the breaking of the chorion and vitelline membrane, and it may be that it is simply the emerging from these coverings of the first instar. Alternatively, the little arachnid may shed its cuticle once or even twice while still in the egg, and emerge in its second or third instar. These differences follow familial or generic taxa rather than specific or individual instances. One of the characteristics of the class is that the first free-living form is at once recognizable as an arachnid, and is not, in the generally accepted sense of the word to be described as a larva, but rather as a nymph. Invertebrates that share this resemblance between the adult and newly hatched young are said to show "direct development". Customarily the nymph is described as differing from the adult only in its smaller size and its inability to reproduce, but among Arachnida careful and detailed study of the younger stages has revealed organized, steady progress in the course of growth. The young arachnid passes through periods that can be precisely defined and recognized.

For Araneae, these have been named by Vachon and Legendre as:

(1) Embryonic Period: from fertilization of the ovum to the end of inversion of the embryo.
(2) Larval Period: from inversion to the appearance of an active animal capable of an independent life.
(3) Nympho-imaginal Period: from leaving the cocoon until death.

Each of these main periods may be further analysed:

Embryonic:
(i) Primordial Phase: from fertilization to the formation of the germinal layers and germinal disc.
(ii) Metamerism Phase: from inversion to a segmented embryo, of an elongated form and with appendages beginning to appear.

    (iii) Inversion Phase: the curvature of the embryo is reversed and a typical arachnid form is assumed.

Larval:
    (i) Pre-larval Phase: the embryo is immobile, the metameres all but invisible, the sense organs absent.
    (ii) Larval Phase: spines, setae and tarsal claws appear.

Nympho-imaginal:
    (i) Nymphal Stadia: 1 to x. The stadia are separated by ecdyses, the number of these determined by the adult size.
    (ii) Adult Stadium: sex organs mature and reproduction is possible. An additional stadium occurs in species that undergo a post-nuptial moult.

In an examination of this summary, the introduction of the term larva is to be noted, for the arachnid larva differs from the larvae of Insecta and Crustacea in that it does not lead an active life, feeding itself and dispersing the species before undergoing a metamorphosis. The larval spider is the object that breaks through the blastodermic cuticle. Although it looks like a spider, its prosoma and opisthosoma are at right angles to one another, it has no sense organs, and it is unable to move, feed or spin. This condition may persist through two moults, so that a distinction has been drawn between the "pre-larva" and the larva. The third moult brings about so many changes that it almost justifies the description of a metamorphosis. It produces a nymph which can move, feed and spin. The nymphal condition lasts through a number of stadia, which varies with the size to which the spider will grow. There may be 12 nymphal instars, the last of which may be visibly distinguished, and is always distinguished in males, as the sub-adult instar. Nymphs resemble adults in all essentials except size, the possession of secondary sexual characters and the ability to reproduce.

In general, it may therefore be said that the so-called larval period is a stage of transition between embryo and nymph, during which the organ systems are perfected. It is more in keeping with general nomenclature to call it a prolongation of the embryonic state.

Araneists may well be advised to abandon their traditional habit of neglecting or even throwing away the immature specimens that they find, for these have much to teach us, and even the cast-off exoskeletons left after moulting may be profitably examined. The successive nymphal stadia of any one species are not readily distinguishable, but one example of the gradual bodily changes is given by Vachon's investigation of the regeneration of autotomized legs.

He was able to distinguish six types of legs, differing in their trichobothria, in the number of segments, and in the existence of smooth or toothed claws. These are shown at Fig. 19. He has called this arthrogenesis, and it is obviously an aspect of growth to which more attention might well be given.

The above considerations are applicable to the growth stages of pseudoscorpions, as described by Gabbutt and Vachon (1967, 1968). As was described in the previous chapter, the false scorpion still in the brood-sac is fed by its mother until it moults, and at this stage almost justifies the description of larva given it by Vachon. Then it moults

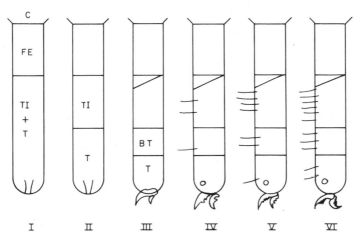

Fig. 19. Arthrogenesis of a spider's leg. After Vachon. (I and II) Pre-larval stage; (III) Larval stage; (IV, V and VI) Larval stage.

and escapes into an independent existence as a protonymph; and by successive ecdyses deutonymph, tritonymph and adult follow. In the few species that have been intensively studied these four instars can be distinguished in a number of ways, such as micro-measurement of their proportional dimensions.

Features which, among others, provide the most readily perceived distinctions between the stadia are the numbers of setae on the chelicerae and carapace. On the chelicerae the fixed and movable fingers are taken separately; on the carapace four rows of setae are named anterior, ocular, median and posterior. The authors named have given the following figures for two common European species:

RONCUS LUBRICUS

	Protonymph	Deutonymph	Tritonymph	Adult Male	Adult Female
CHELICERA					
Movable	0	1	1	1	1
Fixed	4	5	6	6	6
CARAPACE					
Anterior	4	4	4–5	4	4
Ocular	4	4	5–7	5–6	5–7
Median	6	8	7–8	8	7–8
Posterior	4	6	6	6	6

MICROCREAGRIS CAMBRIDGEI

	Protonymph	Deutonymph	Tritonymph	Adult Male	Adult Female
CHELICERA					
Movable	0	1	1	1	1
Fixed	4	5	6	6	6
CARAPACE					
Anterior	4	4	4	4–5	4
Ocular	4	4	4	4	4
Median	4	6	6	6–7	6
Posterior	4	6	6	6–7	6

Young harvestmen are feeble little creatures, which stagger about on their long legs in a most unimpressive manner. The tarsi have fewer articulations than in the adult state: thus, recently hatched Oligolophus has eight, 18, eight and eight pieces to the tarsi of its four legs compared with the 26 to 50 of the adult. A more remarkable contrast is found in the chelicerae. In the adult the forceps are smooth or nearly so, but in the very young they are conspicuously or even strongly toothed (Fig. 20), as if this additional armature were a compensation for small size and bodily weakness.

A newly-hatched solifuge (Fig. 21) is an immobile and incomplete creature, existing in a quasi-larval condition until its first ecdysis changes it into a protonymph. This has a smooth carapace, on which two transverse furrows foreshadow its later division; and by comparison with an adult its opisthosoma is also to be described as smooth, for it carries but six or ten pairs of dorsal setae. The movable finger of the chelicera has a small egg-tooth; the pedipalpi and first legs are directed backward, lying across the other legs. All legs are smooth, their seg-

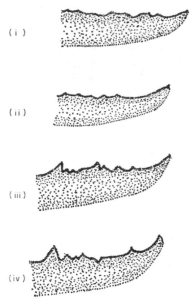

(i)

(ii)

(iii)

(iv)

FIG. 20. Chelicerae of very young harvestmen. (i) Phalangium; (ii) Homalenotus;
(iii) Mitopus; (iv) Oligolophus.

ments ill-defined or invisible. Claws are present, and a very remarkable
racket-organ is situated behind the first and second legs. There are
various, small, external differences between the "larvae" of Galeodes
and Solpuga, but internally they agree in having a blindly ending
alimentary canal, in which there metabolizes a mass of yolk, providing
both energy and material in the apparent absence of functioning respiratory and excretory systems. There is clearly much to be learnt about
the physiology of all very young Arachnida.

Knowledge of the life cycle of the Uropygi has been provided by
Weygoldt (1971), who described the life of Mastigoproctus. In a closed
burrow the female produces from 20 to 40 eggs, delivered into a brood-
sac which remains attached to her body. Each egg develops into a
pre-nymph, a whitish organism about 12 mm long. After an ecdysis the
pre-nymph becomes a true nymph, leading an independent life: a sucker
at the end of each leg enables it to climb upon its mother's back. A
month or so later another moult makes them definitely recognizable as
Uropygi, but they continue to live in the burrow with their mother and
to share her food. Four more ecdyses, occupying four years, are needed
to reach maturity. Adults may live for as long as 14 years.

The life story of the Amblypygi is essentially a similar one.

The life cycle of the Ricinulei is well illustrated by Mitchell's account

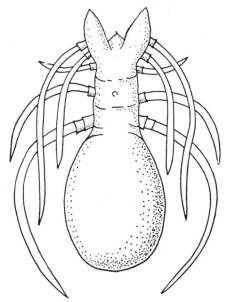

Fig. 21. Newly hatched, or post-embryo, solifuge.

(1972) of the species *Cryptocellus pelaezi*. From the egg there emerges a larva, characterized by the possession of only three pairs of legs. The tergites of its opisthosoma are well separated and the tubercles on the carapace and elsewhere are scattered without arrangement. The larva is followed by three nymphal stages, the first of which differs from the larva mainly in the appearance of the fourth pair of legs. The second and later nymphs bear a longitudinal furrow on the carapace, and the tubercles begin to take on a definite arrangement. The third nymph continues this arranging, while the longitudinal furrows on carapace and abdomen become more marked. All stages up to this are cream-coloured, but the adults are reddish-brown.

The five stages differ in the number of segments into which the tarsi of the legs are subdivided. These are as follows:

	Leg 1	Leg 2	Leg 3	Leg 4
Larva	1	2	2	2
First nymph	1	4	3	2
Second nymph	1	5	4	4
Third nymph	1	5	4	5
Adult	1	5	4	5

No arachnid life history is more remarkable than that of the Cyph-ophthalmi, which lasts for nine years. This is described in Chapter 25.

The period of actual growth in the true sense of increase in size is confined to a short interval, probably not more than a couple of hours. following the moulting or casting of the exoskeleton. The ecdysis, as it is properly called, is not difficult to forecast in spiders, for their legs change colour a day or two before the moult, darkening until they are almost black. The spider also ceases to feed and its silk glands are affected, so that the fluid silk within becomes more viscous and the spider scarcely spins a thread.

Among harvestmen there are in general no such warnings, but false scorpions show a change of behaviour. They wander about, as if they were looking for a suitable place, and when they have found one they build a protective cocoon or moulting chamber.

In all the orders the operation is the same in its essentials. A moulting fluid appears to be secreted between the old exoskeleton and the new one, acting in part as a lubricant and in part as the cause of the pressure which produces the first split. This generally runs round the sides of the prosoma and along the middle of the lower surface of the opisthosoma. The carapace easily falls off and the cuticle of the abdomen shrivels. The difficult part of the business in all orders is the extraction of the eight legs from their old cases. For a spider this may take anything up to half-an-hour or more, as the legs are slowly pulled out by a series of rhythmical heaves. Among harvestmen, whose legs are longer, the extraction is assisted by the chelicerae, which grasp the legs and pull them.

Araneomorph spiders normally hang themselves up on silk threads from the spinnerets during the process of ecdysis. Harvestmen do not produce silk of their own, but they usually moult in a suspended position, with the legs of the fourth pair hooked on to some suitable support. Their skeletons may sometimes be found on walls, hanging on a thread of spider's silk; but in the absence of such means of support they are able to moult, like false scorpions, standing up.

In any account of ecdysis the obvious questions which suggest them-selves are: how long passes between one moult and the next, and how many times does an arachnid moult?

Bonnet (1930) has shown that the number of ecdyses that a spider makes depends on the size that it is destined to reach. There may be only four moults for a spider under 5 mm long, seven or eight for one twice as long, and ten or 12 for a larger one still. At the extremes stand Gertsch's observation that a male *Mastophora cornigera* moults twice only, and Baerg's record of 22 for *Eurypelma californica*.

Further, the number is not constant, even within a species, for it is affected both by sex and by food supply. Male spiders, being smaller, usually moult less often; underfed spiders moult more often. Moulting does not always cease when a spider is mature; post-nuptial moults occur annually or more often among spiders with long lives.

After an ecdysis has been successfully carried through, a spider may often be seen to stretch, bend and straighten its legs continuously for some time. It is during these minutes that the new exoskeleton hardens, and these exercises prevent the hardening process from spreading across the joints and rendering the limbs rigid. The preening, which takes place intermittently at the same time arranges the spines and setae and causes them to set in the correct directions.

Less is known about the number of ecdyses among Arachnida of the other orders. The scorpion *Buthus occitanus* has been known to moult seven times, and Guenthal (1944) has recorded eight moults for the harvestmen *Phalangium opilio*. This was accomplished in 30 days at a temperature of 30°C, which considerably hastens the rate of development. At ordinary temperatures about 10 days pass between one moult and the next.

False scorpions have been found by Vachon (1933) to undergo ecdysis four times. The first moult of the species *Chelifer cancroides* occurs in the incubation chamber, when the egg hatches into the "larva", and this as mentioned above proceeds in turn to a protonymph, deutonymph, tritonymph and adult.

Muma (1966) has recorded eight or nine ecdyses in the life of the solifuge Eremobates.

A most important aspect of this process of ecdysis is the regeneration, or reproduction of lost limbs, that accompanies it. Nearly all Arachnida have the power to cast an appendage if it is injured or seized by a predator, but the loss is only a temporary one. A new limb forms beneath the exoskeleton, and comes into use when the next ecdysis occurs. This power of regeneration is not shown by harvestmen, which, in fact, drop their legs with great ease, but it is known to exist among scorpions, false scorpions and spiders, and has been exhaustively studied by Bonnet (1930), working with the spider *Dolomedes plantarius*. His conclusions were that lost limbs require three moults before they attain their normal size, a fact which explains the asymmetry often to be seen in spider's legs. More than one leg can be regenerated at a time; in fact Bonnet pushed his investigations to the limit by keeping several spiders which had lost all eight legs. Their almost immobile bodies were carefully tended and supplied with water to drink. They accepted flies offered to them in forceps, and ate as much as normal specimens. When the time of ecdysis arrived, all eight legs reappeared.

The character of the new limbs has been studied in great detail by Vachon (1941). As was mentioned earlier in this chapter, he distinquishes between embryonic, pre-larval, larval and nymphal instars, and among the distinguishing characters are the leg spines and the claws. Thus Vachon discovered that if a spider loses a leg just after a moult and later loses the corresponding leg just before the next moult, there will appear at that moulting for the first loss a larval leg and for the second an embryonic leg. At later ecdyses, the regenerated legs go through the recognized stages of development, but at an accelerated speed, one stage being missed out; and it is also found that in the same regenerated leg the segments are not necessarily all at the same stage, the distal ones being "older".

Length of life is very variable among Arachnida. Fundamentally four types of longevity can be recognized.

(i) The commonest is the life cycle of nearly all the small species, and occupies rather less than a year. Eggs are laid in the autumn, the winter is passed in development, and the young hatch in the spring. They grow up during the summer, are mature in early autumn, mate, lay their eggs, and die before the winter.

(ii) A life cycle of more than a year is shown by Arachnida whose eggs are laid in the summer and are hatched in a few weeks. The nymphs pass the winter in hibernation; they mature in the spring, and, having laid their eggs in the summer, they die in the autumn.

(iii) Some Arachnida, including many of the largest, live for 2 to 5 years, having taken from 1 to 3 years to reach maturity. It will be understood that most of the records of such lives as these have come from specimens kept in the safety of the laboratory, where they found equable temperatures, sufficient food, and freedom from their enemies.

(iv) Lastly, there are some, chiefly among the great mygalomorph spiders, which cast their skins annually and in favourable circumstances may live for 15, 20 or even 25 years.

Size, however, is not a determining factor. Juberthie (1967a) has shown that the life cycle among the Cyphophthalmi may last for as long as nine years.

One characteristic of the life span of an arachnid is its dependence on circumstances. Specimens that are kept cool live longer than those that are kept warm; and specimens that are underfed live longer than those whose food is plentiful. It seems clear that a greater metabolic rate wears out the body to a significant extent, and brings its life to a quicker ending.

Abnormalities among Arachnida are no less frequent than among other animals. A two-tailed scorpion was described in 1825, and examples of this continue to be reported. In 1917 Brauer examined 5,000 embryonic scorpions, and found evidence of double tails in 13 of them, as well as duplication of the posterior mesomatic somites.

Pseudoscorpions are sometimes found with irregular tergites. Among spiders abnormal eye patterns are far from unusual, and almost everyone who has collected assiduously has come across specimens in which two eyes have merged into one, or in which one, two, or more eyes are missing.

A very surprising example of unusual development has been shown to follow the submission of the eggs to the action of a centrifuge. Sekiguichi (1957) found that eggs thus treated for 20 minutes at 3,000 revolutions per minute showed duplication of the embryo. An example of one of his results is shown in Fig. 22.

FIG. 22. Duplicated embryo. After Sekiguichi.

Juberthie (1963) pointed out that among the Opiliones abnormalities are rarely found in nature. He illustrated a single case of crossed tergites among Cyphophthalmi found in a specimen of *Siro rubens*, and commented on the fact that other recorded instances of teratology were limited among Opiliones to six species.

He showed later (1968) that monstrosities were often produced if harvestmen's eggs were caused to develop at high temperatures. Eggs of *Odiellus gallicus* kept at 23–30°C produced cases of abnormal tergites, fused limbs, single eyes and an absence of eyes.

Among the most fully described examples of abnormality are the appearances of gynandromorphs. The phenomenon of gynandromorphism is known among other invertebrates, and may be traced to its cause in the behaviour of the chromosomes. The number of gynandromorphs so far described barely amounts to 40 specimens, a much smaller number than has been found in some insect groups, and all but a very few of them have been spiders. Six types of gynandromorphism are possible, and all are known: they can be represented thus:

$$(1)\ \text{M} \mid \text{F} \qquad (2)\ \text{F} \mid \text{M} \qquad (3)\ \dfrac{\text{M}}{\text{F}} \qquad (4)\ \dfrac{\text{F}}{\text{M}} \qquad (5)\ \dfrac{\text{M} \mid \text{F}}{\text{F} \mid \text{M}} \qquad (6)\ \dfrac{\text{F} \mid \text{M}}{\text{M} \mid \text{F}}$$

lateral   transverse   crossed

Since sex is determined by the presence of one or more X-chromosomes in the male gamete, it is reasonable to suppose that an accident to, or a loss of, one of these chromosomes in the earliest stages of segmentation is responsible for abnormal development.

In addition, a small number of intersex specimens have been found. Their peculiarity seems to be due to infection with the nematode Mermis, which partially destroys the gonads.

7

Bionomics: General Habits

Every animal is constrained throughout its life to protect itself from unfavourable circumstances and from living enemies, to feed itself, and finally to reproduce itself; and among Arachnida the ways in which success is achieved in this triple problem are no less diverse than among the animals of any other class.

The dependence of an invertebrate on the physical conditions of its immediate surroundings is always manifest. Fundamentally it underlies much of that side of biology known as ecology, and as such is discussed in Chapter 9. When, however, an arachnid has reached an inhabitable neighbourhood and has discovered an area in which it can remain, it must be able to continue to respond appropriately to all the stimuli to which it will be subject. In other words, it depends on maintaining the efficiency of its sense organs, whether they be eyes, setae or chemoreceptors.

Upon this follows the always surprising fact that Arachnida constantly preen themselves, cleansing and at the same time anointing their cuticles with products of their own making.

The preening of spiders is the easiest of these actions to see, for a common house spider can be made to exhibit them by dropping it into a basin of water. It struggles and, in effect, swims to the side of the vessel, and if it is then picked up and put back in its web, it dries itself very thoroughly.

The legs of the second and third pairs are pulled slowly through the opening and shutting chelicerae, and finished by sucking the water off the tarsus. The pedipalpi are dried in the same way. The first and fourth pair of legs are cleaned a little by this method, but mostly by rubbing against the second or third pairs. The sternum and abdomen are rubbed with the terminal segments of a leg. The separate actions do not take place in any orderly fashion; a little of one is followed by a little of another, and often two are simultaneous. The spider dodges from limb to limb and from side to side without sequence, and the whole operation may take half-an-hour.

The drying of the limbs and body may not be the sole purpose or result of these actions. Holding the tip of a leg in the chelicerae may allow secretions from the maxillary glands in the gnathobases to collect on the tarsus, which then spreads the fluid over abdomen, sternum and elsewhere.

This alternative view is supported by Shulov's full account of comparable behaviour in scorpions. He has described how a scorpion puts the end of a pedipalp between its chelicerae, and adds, "At the same time a liquid, which probably comes from the alimentary canal, is secreted on the pincer". The legs pick up this liquid from the pedipalpi and spread it over the lower surfaces of the body. The end of the tail is then placed between the chelicerae and similarly coated with liquid—Shulov calls it varnish—which it then smears over the carapace and mesosoma. The process may take more than an hour; it may be repeated several times, and it very often occurs when the animal has been drinking or has been wet.

False scorpions show a peculiar modification of the same habit, which can often be seen in captive specimens, and has been fully described by Gilbert (1951). He records the fact that a false scorpion starts to brush its chelicerae and palpi against one another when the prey is observed: the victim may be seized in a pedipalp while the other pedipalp is being passed through the chelicerae. This has the appearance of cleaning the pedipalp, but in actual fact it is the pedipalp which is cleaning the chelicera. It has been observed that the pedipalp may often be allowed to remain dirty, but the cleaning of the chelicerae is in agreement with the actions of spiders when engaged in the same way.

Harvestmen also pass their long legs through their chelicerae at frequent intervals, and the spectacle is the more remarkable because of the length of the leg. By the time that the tarsus has reached the jaws the rest of the leg is bowed into an almost complete circle, and when released it shoots out like a spring. A most significant feature of the preening of harvestmen is the fact that it is almost continuous when the animal is warmed, as when they cannot escape from the rays of the sun.

For Amblypygi this cleaning of the long antenniform legs also appears to be difficult, but it is important because of the animals' dependence on the trichobothria which these limbs carry. The sub-segmentation of the tibia is a helpful adaptation, making it possible so to bend this joint that the leg can be passed to its tip between the chelicerae.

Excessive heat is a constant menace to all Arachnida, partly because it tends to increase the rate of water loss and partly because the firmness of the exoskeleton prevents expansion of the internal fluids, so that vessels and ducts are fatally closed by pressure. This is normally avoided by the adoption of a cryptozoic life, so that the animal, under stones, leaves or logs, is always in the shade and is active only at night.

Scorpions, however, have been shown by Alexander (1958) to meet a rise of temperature by a remarkable habit which she has described as stilting. The scorpion straightens its legs, so that its body is raised above the ground and free circulation of air around it is made possible. Experiment made it clear that heat is the stimulus for this action and that control of the body temperature does, in fact, follow it.

A widespread method of protection and of escape from predatory enemies, which is common among both Arachnida and Crustacea, is the quick dropping of a captured limb. This permits the animal to escape, purchasing life and freedom at the temporary expense of a limb. The habit was closely studied by Wood (1926), who divided the circumstances in which it occurs into the following.

(i) Autotomy is the reflex self-mutilation or automatic severance of a limb from the body. This does not exist in the Arachnida, but is found among Crustacea.

(ii) Autospasy is the casting of a limb when pulled by some outside agent, like the forceps of an investigator.

(iii) Autotilly is the pulling off of a limb by the animal itself, as when an injured leg is seized in the chelicerae and severed from the body.

(iv) Autophagy is the act of eating a part of the body after severance from the rest.

In addition to these a fifth term, autosalizy, has been coined by Pieron (1924) with much the same meaning as autotilly, but apparently involving an element of choice or decision on the part of the animal. According to most views of arachnid behaviour, this phenomenon could not be exhibited by them. It is impossible to imagine that the arachnid, as it pulls on its imprisoned leg, shall come to a decision to run away on seven legs rather than perish.

Wood's experiments showed conclusively that the shedding of a limb is not the result of a reflex action or of a special mechanism, as had previously been believed. It is simply the inability of the skeletal and muscular components of the leg to resist more than a certain force. The leg, when pulled, parts at its weakest point. Consequently, all spiders cast their legs at the coxa-trochanter joint, all "Pedipalpi" at the patella-tibia joint and Opiliones at the trochanter-femur joint. On the other hand, in the scorpions, king-crabs, and some of the mites there is no predetermined locus of fracture. The muscles are evenly arranged throughout the legs and the chitin is well provided with longitudinal fibres at all the interarticular membranes. These factors, combined with a lack of response on stimulation, make autotomy and autospasy impossible and autotilly very unlikely.

The spider *Tidarren fordum*, however, provides a remarkable example of autotilly. The palpal organ of the mature male is large; it is almost half the size of the spider's small body, and Chamberlin and Ivie (1933) found that while all young specimens of both sexes had two palpi, the mature and sub-adult males had all lost either the right or the left palpus. Bonnet (1935) subsequently discovered that during the last instar the sex organs inside the palp cause it to swell to a considerable size. To the small owner one such organ was an inconvenience, but two were intolerable, and the spider therefore pulled off, with its chelicerae, one of its overgrown appendages. It seemed to be a matter of indifference whether the right or the left was removed, but the self-mutilation was invariable. After the last ecdysis the remaining palpus was fully developed and functional. It is clear that this autotilly is habitual in this species and the loss of one palp is not the result of an accident.

A method of defence that is interesting because it is unusual is adopted by several species of the spider family Theraphosidae. Specialized "hairs" of four different kinds grow on the dorsal surface of the abdomen and from here they can be scraped off in numbers by the metatarsal spines of the spider's fourth legs. Under threat a shower of these hairs is directed towards the predator. They produce intense irritation of the human skin, and seem to be used chiefly against small mammals (Cooke, 1974).

Among the Uropygi defence by secretion of a nauseous vapour is more fully developed than among the familiar Opiliones. Glands at the base of the flagellum can produce a spray of acidic droplets, which, owing to the mobility of the abdomen can be directed over a wide range. The method is said to be most effective against birds and mammals.

An arachnid's constant occupation, like that of every other animal, is to provide itself with food.

The average arachnid, if there is such a thing, is a predator, and the picture that rises in the mind as we think of its feeding is one of a nocturnal wanderer, which does not so much seek its prey as come upon it by chance, detecting it by the sense of touch and seizing it forthwith. No doubt this is broadly accurate for many and there is little in such casual collecting that is worth attention. It seems reasonable to regard it as the primitive method of obtaining food, to be contrasted with more specialized methods evolved later. The most obvious of these improvements is perhaps the acquisition of greater speed and strength, or, as in spiders, the highly specialized and unique method of web spinning.

Scorpions seize nothing but living prey, and are well known to eat spiders freely and, more surprisingly, consume large numbers of termites. Frequently the scorpion is attracted towards its victim because the movements of the latter set up air waves to which the trichobothria on the scorpion's pedipalpi respond. The popular picture of the vicious scorpion, formidably armed and scouring the desert, bringing sudden death in its path, must be modified in the light of observed facts. Scorpions seldom persist in attacking an adversary that puts up a spirited defence, and they are unable to injure anything which, like a large beetle, is protected by hard chitin. They do not often eat each other; even in captivity they are seldom cannibals if alternative food is provided. When they have made a capture they rest for a long time while they consume it; and their method of lying in wait is well developed. Berland (1948) has described the course of events that follow when a cockroach is put into a scorpion's cage. The latter does not take a step, but remains "confident in the ascendancy which its immobility gives it". The agitated cockroach rushes about and before long it comes within reach of the scorpion's claws. That is enough.

Solifugae are more active and are also more rapid consumers of their food, which includes many insects, with termites among them, as well as bed bugs, which they have been said to relish. When captive Solifugae have been offered crickets, they have been known to consume as many as nine a day. As soon as one of the insects came into contact with one of the long setae on the solifuge's body, there was a short spring forwards, and a lightning snap of the great jaws. The fact that a solifuge is eating a cricket will not deter it from seizing another to follow it; and the crunching of the victim's body can be heard at a distance of several feet.

This vigorous mastication by Solifugae is very characteristic of the order. The two scissor-like chelicerae hold the body horizontally, and as they bite it, they also move backwards and forwards alternately. No other arachnid eats like this; it is one of the characteristic features of these animals. Their long leg-like pedipalpi are so important as tactile organs that the solifuge does not long survive their loss; in this respect these limbs resemble the equally essential second legs of harvestmen; and Solifugae also resemble harvestmen in their constant need for water and their omnivorous habits.

Harvestmen come upon their prey by chance during their nightly wanderings, and almost alone among Arachnida they are frequent feeders on dead matter. Their nocturnal habits have not deterred the enthusiasm of either Bristowe (1949) or Sankey (1949), who have made expeditions with the help of an electric torch to discover exactly what harvestmen do at night. They independently noticed harvestmen eating other harvestmen, snails, worms, millipedes, woodlice, earwigs and flies, all of which the harvestmen had apparently captured and either killed or disabled. They were also seen to be eating the bodies of dead ants and beetles, and Bristowe observed the nibbling of the gills of fungi. Other observers have told how they have been seen feeding on dead moles or mice and sucking the juices from bruised and fallen fruit.

An account of the harvestman's method of catching its prey has been given by Roters (1944), who fed his harvestmen chiefly on flies and small moths. He wrote that the moths were pursued as soon as they were put into the cage, the harvestmen reacting at once and beginning to search, tapping the ground with their second legs. When an insect was touched by one of these legs a swift dart forward surrounded the victim with a screen or palisade of legs, while the pedipalpi sought to make contact with it. If the pedipalpi touched one of the wings, this wing was seized in the chelicerae and passed through them until the body was reached. If the imprisoned moth moved, the harvestman raised its body aloft and suddenly dropped down upon it like a living pile-driver. The body of the moth was devoured while the wings were held down by the harvestman's legs.

Roters' general conclusion was that the choice of prey is less a matter of taste than of ability to capture it.

If a harvestman is watched while it is feeding, it can be seen that the food is held in place by the pedipalpi and chelicerae. The forceps of the latter tear at the skin until it splits, when one pair of forceps is thrust into the wound, drags out what it can grasp and carries the "handful" to the mouth. The blades on the coxal segments of the pedipalpi and the first pair of legs then open and engulf it. The blades of the second pair of legs, which in many species are immovable, seem to take no part in the process, while the function of the lip seems to be to prevent the loss of fluid from the oral cavity.

A harvestman's feeding is thus different from that of a spider. It is not limited to fluids which can be pressed or sucked from the prey, nor is the animal so accustomed to a large meal following, or followed by, a long fast.

False scorpions resemble spiders and true scorpions in confining themselves to victims which they have just caught and killed. Cannibalism is occasional, but is infrequent if other food is available. The chief diet of the European species seems to be found in the many small insects, such as spring-tails, which live in the same environment. A study of the feeding habits of several species has been made by Gilbert (1951). He recorded the way in which a false scorpion will start to clean its chelicerae with its pedipalpi when the prey is observed, and says that the victim may be seized either by pedipalpi or chelicerae.

Sometimes the prey was held by one pedipalp while the chelicerae were being cleaned by the other.

The food particle was torn open by the chelicerae, which were thrust into the wound, where they sometimes remained stationary and were sometimes moved about. The flooding of the food with secretion from the cheliceral glands was clearly seen. The duration of a meal was from 1 to 2 hours and was followed by cleaning of the rostral area. There are differences between the processes of feeding by different species, depending in part on the presence or absence of venom in the pedipalpi, and in part on the size and structure of the chelicerae.

Only a slight familiarity with the ways of false scorpions is sufficient to impress one with the outstandingly potent nature of their venom. A victim once bitten seems to be at once immobilized; and one has seen a spider, swinging on its thread, and coming into momentary contact with a false scorpion much smaller than itself, die instantaneously when it was bitten. It is tempting to claim for false scorpions that in proportion to their size, they are the most venomous of all the Arachnida.

Two topics complete a review of an arachnid's nourishment. The first is the related problem of drinking.

There is a common belief that scorpions do not drink, and even a fable that water kills them, but neither statement is true. There are several species of scorpion that can live only in damp places, where drinking water is always available. In this they resemble harvestmen, which drink frequently and, like false scorpions, soon die if the atmosphere is too dry. On the other hand, there are scorpions and harvestmen that live in hot deserts, and the former at least are known to be able to survive because of a well-developed power of water conservation, due to a wax layer in the cuticle.

The second point is the comparative indifference of many Arachnida to fasting. Captive Arachnida do not need to be fed every day, and many a spider is overfed if given a fly daily. A couple of weeks without food seems to be of no importance whatever, save perhaps to harvestmen, which do not appear to be blessed with quite the same indifference. But this is nothing when compared with the feats of survival that have been recorded by various observers. Iconopoulous kept a scorpion without food for 14 months, Jacquet kept one for 368 days. Among spiders, Berland had a Filistata that did not feed for 26 months and Blackwall a Theridion whose record was 30 months. Baerg has kept a trap-door spider for 28 months.

Although this has been known for so long, precise investigation of the spider's condition during a period of abstinence has but recently been carried out. Anderson (1974) kept a number of spiders of two species,

Lycosa lenta and *Filistata hibernalis*, the normal life spans of which species are ten months and several years respectively. During the experiment the spiders were weighed and measured every fortnight and their metabolic rates were determined by measurements of their oxygen consumption.

Kept without food, the Lycosas survived for an average of 208 days and the Filistatas for 276 days. Their metabolic rates were reduced by 30 or 40%, with no apparent decrease in their ordinary capabilities; and if the starving were stopped they quickly doubled their weight.

The most significant result was that differences in survival time were inversely proportional to the metabolic rates, an adaptation which increases the individual's chance of coming through a time of natural scarcity.

The reproductive habits and activities of Arachnida are no less characteristic than their methods of protecting and feeding themselves. In all species the sexes are separate and the ova are fertilized internally, usually by spermatozoa which have been stored in spermathecae and which enter the ova at the time of laying. The transference of spermatozoa from the male to the spermathecae of the female is seldom a simple process; it is preceded by a series of actions, sometimes prolonged and complicated, which are customarily, if unfortunately, described as courtship.

This courtship is best known and is perhaps most diverse among spiders, in which order it takes a wide variety of forms. The vagabond species of more primitive families usually show little more than a mutual touching and stroking, after which the male climbs on the back of the female and inserts into the epigyne a palpal organ which has been charged with semen. More elaborate preliminaries are shown by spiders with better eyesight, like those of the families Lycosidae and Salticidae. The male indulges in a kind of dance: secondary sexual differences such as black segments on the legs or coloured patches on the body, are displayed before the female, who appears to watch them as the dance proceeds. This kind of courtship is more demonstrative among jumping-spiders than among wolf spiders, and many descriptions of individual performances have been written and illustrated.

Web spiders do not dance in this way. The male, who with maturity has abandoned his web and has thereby sacrificed all chance of another meal, arrives during his wanderings on the outskirts of a web occupied by a female of his species. His reaction consists either of drumming vigorously on the web with his palpi or in tweaking or twitching its threads in some way which does not stimulate the female to rush out as to a captured fly. In this way he makes a gradual approach until his forelegs touch hers.

This initial contact is evidently a critical point in the meeting of any two Arachnida in many circumstances. It seems to act as a stimulus to the curious behaviour first described among harvestmen as "combats acharnés" (Simon, 1879). These bloodless battles have also been seen by Weygoldt (1971) as occurring between individuals of Amblypygi and Uropygi. After the first touch the two animals meet with widely separated pedipalpi, each one pushing and pulling the other. After a few seconds all is over, one of the two retreats slowly and is disinclined for a second round. That there is no enmity in these displays is shown by the fact that if many specimens are together in a cage with no room to manoeuvre there are no battles and all remain peaceably together.

The courtship behaviour found in other orders is not so colourful, but it is just as unexpected. It has been seen among scorpions, false scorpions, Amblypygi and Schizomida, and an unmistakable similarity runs through the performance in all these groups. There are specific differences, and no doubt when more species have been observed more often, these courtships will be recognized as constant and as individualized as they are among spiders. In a shortened and generalized outline the course of events is as follows. The male grasps the female with pedipalpi or chelicerae, and thus united they walk to and fro, at times varying the routine by some change of position, until, perhaps after an hour or more, the male secretes a spermatophore (Fig 23).

Fig. 23. Spermatophore of scorpion, Opisthothalmus. After Alexander.

This is an object produced in gelatinous form from his genital orifice; it adheres to the ground and quickly solidifies. On top it carries a drop of semen. The result of the dance has been to bring the female into such a position that if she is led forwards the spermatophore will enter her epigyne. This the male does, guiding her accurately, and among false scorpions then shaking her so that the spermatozoa are detached from the top of the spermatophore and liberated inside her body.

Both the dances of spiders and the promenades of scorpions are surprising actions. Clearly instinctive in nature, they ask for an explanation of a kind that will help us to understand why the vital fertilization of the ova should require such preliminaries and suffer such delay. Any explanation must take into account the habits of two large orders which have not yet been mentioned, Solifugae and Opiliones.

Among Opiliones there is no courtship and the sexes unite as soon as two mature individuals meet. The male possesses a chitinous extrusible penis (Fig. 24), which passes between the chelicerae of the female as the two animals face each other, and liberates the spermatozoa in a few seconds. Solifugae show neither display nor dance; the male "caresses" the female with his legs and palpi, as a consequence of which she falls into a state of catalepsy or trance. While she is thus unconscious he picks up a globule of semen in his chelicerae and presses it into her genital orifice.

Clearly, among Arachnida, courtship is a many-sided phenomenon, and any explanation of it must not only cover its diversities but should

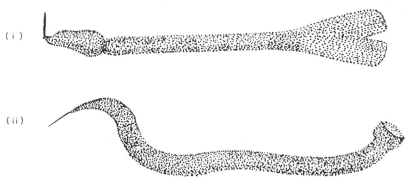

(i)

(ii)

FIG. 24. Male organs of two harvestmen: (i) Oligolophus and (ii) Nemastoma.

also be applicable to other classes of invertebrates, such as insects and molluscs, which exhibit parallel forms of action. An attempt to do this is included in Part V.

Behaviour that may be described as peculiar to motherhood is found sporadically among Arachnida. Many leave their egg-cocoons to their fate, and are themselves dead before their young are hatched. The exceptions are chiefly to be found among the spiders that carry their cocoons about, either in their chelicerae or clasped under the sternum by pedipalpi and legs or attached by a short silk thread to the spinnerets. The last of these, shown by all wolf spiders of the family Lycosidae is the most familiar, and is a habit shared by only a few others.

Among the Lycosidae the cares of motherhood are extended to the carrying of her young on her back for a time after they have left the cocoon. The habit has been investigated by Rovner *et al.* (1973), who, using a scanning electron microscope, has seen on the upper surface of the abdomen knobbed setae peculiar to female Lycosidae. These afford a firm grasp to the undermost layer of the brood in their precarious situation. The knobbed setae are not innervated, but long, smooth setae are in contact with nerves and probably play a part in determining maternal behaviour.

The characteristic of all these methods is the evident bond between the spider and her cocoon. Although we cannot know whether the cocoon appears to her as an object of affection or of value, we are soon aware of the facts that she resists any attempts to remove it, that if it is taken away by force her behaviour is conspicuously changed, and that if it is offered to her again she at once accepts and replaces it, reacting not to its appearance but to its touch. Investigators have shown that a deprived spider will accept the cocoon of another spider or any one of a wide variety of substitutes, such as a pith ball or a pellet of cotton wool or blotting paper; and in nature such spiders have been found carrying small shells.

The memory of a bereaved spider for her cocoon varies from a few hours to a few days, after which she will accept neither her own cocoon nor any substitute. It is believed that her behaviour is dependent on the state of her ovaries; if the arrival of a new cocoon is near, her memory is short, although she has carried it for longer than if the arrival of the new cocoon was in the distant future.

No male arachnid ever shows any parental responsibility and no male wolf spider has ever been recorded carrying a cocoon.

The time of courtship is but one of the occasions when Arachnida are in communication with one another, and it is more easily recognized by ourselves because the movements involved are unusual and conspicuous. The study of intercommunication in a wider sense is receiving closer attention, and Rovner (1967b) has found that Arachnida can affect one another by drumming or scraping on the ground, by a display of colours, and by particular waving of the legs or palpi. These are additional to the acts of stridulation, found among scorpions, spiders and in other orders.

Several of the sounds made by spiders have been recorded, and playback experiments have proved that spiders can hear air-borne sounds.

8
Ethology: Behaviour

The descriptions of the daily habits of Arachnida given in the last chapter form an example of what may be called the simple or natural history method of studying animal behaviour. They attempt to provide answers to the question: how do these animals behave? They thus supply material on which to base the more valuable discussions which attempt to answer the more difficult question: what makes these animals behave as they do? To do this the different types of activity must be analysed, classified, and their characteristics defined.

The simplest complete reaction of the nervous system of an animal is the form of response known as a reflex—"a neuro-muscular adjustment due to the inherited mechanism of the nervous system". The familiar examples of reflex action, such as the blinking of an eyelid, are often described with an emphasis on the speed with which the whole system carries out its function of appropriate response. Thus there is a tendency to neglect the fact that many important reflexes have to persist for hours at a time, an uninterrupted series of unit actions. Important examples of this are the reflexes of posture or of habitual attitude, the maintenance of which is their chief function. These are called tonic reflexes. The majority of reflex actions are quickly fatigued and quickly recover, but the tonic reflexes of posture do not tire in this way. On the other hand, they are the most easily dispossessed or over-ridden of all, as if they paid for the continuous use of the nerve-paths by an exaggerated readiness to make way for other traffic, and in this we can perceive their nature and function. They form a perpetual substratum of nervous activity, maintaining the body in a state of dynamic equilibrium. It is clearly important that this state of equilibrium shall be upset easily, so that there shall arise that agility of response to the changing circumstances of life which produces efficiency and successful activity in the living animal.

A very large proportion of the life of an arachnid is spent in waiting, and during these long periods of inactivity the tonic reflexes are in sole possession of the nervous system, supporting the body on its eight legs against the unceasing gravitational force. For animals like the harvestman *Leiobunum rotundum*, with its tiny body and eight long symmetrically arranged legs, the tensions in the muscles of all legs are probably almost equal, and for spiders like the Linyphiidae which hang inverted below their hammock-like webs, the tensions are probably not very unequal. It can scarcely be a coincidence that these types are particularly sensitive to surrounding disturbances. In the common garden spider, which hangs head-downwards in its lovely orb-web, the weight must clearly be disproportionately supported by the two hind pairs of legs and the same is true of almost any arachnid that stands on the ground. The heavy opisthosomatic region has no legs of its own and the greater part of its support must come from the posterior pair of thoracic legs. It should not be surprising, therefore, that when the equilibrium of the tonic reflexes is upset, the readiest response is a flexion of the femora, so that the patellas tend to meet over the middle of the animal, and especially a flexion of the forelegs. This simple automatic response, which may be called the flexor reflex, is easily witnessed whenever a spider is, as we would say, frightened, and several valuable consequences follow from its operation.

One of these is the flash-colouring method of protection, which most naturalists would associate with the tree frogs of tropical forests, but which is also well shown by several British spiders. A good example is *Segestria senoculata*, a common species with diamond-shaped marks on its abdomen and bright tawny femora. As it runs, the moving femora are conspicuous, but when it suddenly stops and draws in its legs this brightness vanishes and the observer finds the spider invisible.

Without any help from flash-colours, the flexion assists in producing invisibility in other spiders. If on the beach in certain parts of the coast the heaps of dried seaweed are turned over, there runs out a sombrely-coloured active spider, *Philodromus fallax*. As it runs it may easily be mistaken for a grain of sand, rather bigger than the average, rolling down a slope, and, in fact, such grains of sand are often mistaken for spiders. But there is this difference. When the pellet of sand comes to rest it can be seen; when the spider stops moving it is invisible. It does not bury itself nor leap suddenly aside; it simply stops and flexes its legs. And it is gone. One may stare straight at one of these spiders and be quite unable to distinguish it until it moves again.

This leads naturally to the obvious and probably the most important of all the consequences of the reflex, the action metaphorically described as "feigning death" and less inaccurately as the cataleptic reflex. In many Arachnida a sudden disturbance causes them to draw in their legs and fall motionless, and in some the caresses of the opposite sex produce the same condition. Various considerations make it clear that the animal is not in a true state of catalepsis. Often after first falling motionless it will re-arrange its legs slightly, and in some species a periodical tremor runs through the limbs. In addition to these movements, which would not be seen in true catalepsy, there is not the complete insensibility which would be expected. If the motionless spider be gently touched it will often get up and run away. If the spider is lying low at the end of a thread attached to its web and the web be touched with a vibrating tuning-fork, the spider at once awakes and returns to its web.

The whole action is nothing but a sustaining of the flexor reflex and the more closely it is studied, the more clearly it shows the characteristics of reflexes in general. Many reflexes continue to discharge after the stimulus ceases, and in the "cataleptic" arachnid we are witnessing the retention of the position produced by the flexion of the femoral muscles. All reflexes, however, are subject to fatigue which causes them to give up possession of the nerve-paths in favour of some other impulse, previously inhibited. Thus the quiescent spider awakes and a response to a probably new situation occurs.

In several families of spiders the flexor reflex operates with a different result. The common house spiders, Tegenaria, are familiar examples of spiders which rest in the corner of a sheet-web. Their forelegs are outstretched and their claws grasp the silk so that often the sheet can be seen to be drawn up into small cones. If an insect brushes against the web, the legs are jerked inwards, plucking sharply at the sheet; and the result is not invisibility for the spider but a further entanglement of the insect. Orb-web spiders perform exactly the same action, the web shakes and the spider plucks at it. Very often the spider then turns about and repeats the jerking in another direction, an action which, despite its intelligent appearance, is only a consequence of unequal tensions in the threads of the web. The inequality may be due to the recently added weight of the captive and it is this which turns the spider round, and not an attempt to discover whereabouts in the web the arrival has landed. Like beauty, which lies in the eye of the beholder, purposiveness lies in his interpretation of what he beholds.

A consideration of a second reflex action furnishes support for this view. If a reflex is found to have no manifest result or no meaning in an anthropomorphic sense, the observer refrains from imputing purpose to the animal and is content to describe the action itself as a mystery. Spiders often exhibit one such action. If the flexor muscles raise the femora while the extensor muscles straighten the patella, the result will be that the spider lifts its leg into the air, stretched up at right angles to its body. This is most often seen in orb-web spiders, for which it constitutes the normal reaction to a sudden noise. A whistle, a cough or the bark of a dog near a full-grown Araneus usually makes it shoot out its forelegs, as if reaching towards the origin of the sound. I have recorded an observation made in September 1970, when I was walking round a friend's garden and speaking about the many spiders whose webs we passed. One and all they responded as we spoke, raising their legs in response to our voices.

The anthropomorphic description of these actions would be to say that the spiders were listening but whatever it is that the spider hears it does not proceed to any further action, and the idea that it can hear better with its toes in the air is too foolish to be suggested. Moreover, other spiders perform the same action in quite different circumstances. Some female spiders in charge of their egg-cocoons assume this position when driving off intruders, others similarly repel the advances of an unwelcome male. Many spiders raise their forelegs in courtship, others do so when teased or threatened. It seems reasonable to interpret the otherwise meaningless act by suggesting that it is elicited by a different kind of disturbance from that which produces complete flexion. The femora are not drawn back so far, the tibiae are not folded against them, the posterior pairs of legs are not moved.

Still another consequence of the flexor reflex remains to be considered. In the performance of many mammalian reflexes there is a characteristic known as the refractory phase. This is the periodic recurrence of a condition of inexcitability of the flexor muscles, so that the limb is momentarily straightened by the extensors, and the result is a rhythmically repeated motion as in scratching. Arachnida do not scratch themselves, but some of them in certain circumstances will show a rhythmic repetition of the flexor reflex, which again has valuable results. Two very common spiders, *Araneus diadematus* and *Pholcus phalangioides*, possess the habit of rapidly shaking their webs so that they themselves, usually conspicuous in the centre, become blurred and indistinct. This shaking is achieved by vigorous contractions of the femoral muscles, suitably timed, but even if the stimulus continues the reflex will cease. Reflexes can be stopped either by fatigue or by inhibition from another reflex. Under inhibition the reflex fades out with no change in the frequency or extent of the beats, and this is seen if one touches the oscillating spider, causing it to drop at once. Under fatigue the actions continue with a slower rhythm and this is seen if the spider can be forced, as by gently blowing on it, to continue its vibratory movements.

A very extensive group of actions, which cover a large proportion of an arachnid's activities, are traceable to the symmetrical arrangement of its sense organs. The ocelli, the spines, the setae, and other receptors on the body and limbs give scope for a much more elaborate series of directed responses than can be determined by the mere pairs of two eyes and two ears of the vertebrate. The resulting movements take place in directions which cause symmetrical stimuli on the two sides of the body. They were formerly known as tropisms, a term derived from the botanists; and the tropism theory of animal conduct was energetically developed between 1890 and 1920 by Loeb and his colleagues. It showed the way to much experimental work, which finally led to its replacement by the ideas of kineses and taxes.

Kineses describe the effects of stimuli on the rate of random movements, such as appear to be "trial and error" actions. In affecting the rates and not the directions of movements, kineses are essentially distinguished from taxes. When there is no more than a simple effect on the rate of locomotion, related to the intensity of the stimulus, the behaviour is described as an orthokinesis; when the effect appears as a change in the frequency of turning, or rate of change of direction, the behaviour is described as a klinokinesis.

Trial and error movements have been called phobotaxes; but the name kineses is preferable, because it does not suggest a conscious quest, and is free from the human association of the word phobia.

Taxes, on the other hand, are concerned with orientated movements, depending on the direction from which the stimulus arrives. Normally they depend on the existence of symmetrically placed, paired sense organs, and their effect is seen in the direction which the animal takes when it moves. Kühn (1919) divided these movements into four groups.

    (i) Trophotaxis: the animal directs itself symmetrically.
   (ii) Menotaxis: the animal preserves a fixed direction with respect to the stimulus.
 (iii) Mnemotaxis: movements in which memory plays a part.
 (iv) Telotaxis: movements directed towards a goal.

It is clear that the first group of movements closely resemble the tropisms of Loeb. The distinction is based on the idea that trophotaxis involves sensation, in other words it makes the assumption that the animal has a conscious appreciation of the stimulation it receives, and turns, for example, towards the light because of the sensation of brightness and not because of the chemical changes which it produces. Such a conception cannot explain the fact that the behaviour of certain animals, such as Eudendrium, is in agreement with the physicochemical law which involves the intensity of illumination. The sensation of brightness is not proportional to the intensity of the stimulus, but the concentration of the products of the photochemical reaction is, and the behaviour of many animals agrees with the latter, not the former.

Telotaxis is an orientation of the animal followed by movement towards the source of the stimulus, which is acting as if it were a goal or an end to be sought. Telotactic movements have therefore an outward appearance of purpose, since there is a minimum of deviation in the path taken.

It has been asserted that true telotaxis is shown only by movements towards or away from a source of light, but there can be no doubt that vibrotaxis, which is so conspicuous a part of the behaviour of spiders, is an equally good example. It is readily to be seen whenever a web spider runs across its web to seize its prey. At the same time it should be remembered that when vibrotaxis was first recognized by Barrows (1915) he also described the result of simultaneous oscillation of the web at two foci, when the spider took a path between them. This is typical tropotaxis.

Vibrotaxis is not confined to web spiders, for spiders that hunt on the surface of water respond in the same way to the ripples produced either by struggling insects or by the prong of a vibrating tuning fork. It can readily be seen that this tropism is a direct consequence of the flexor reflex. Let it be supposed that the spider is standing with all eight legs on the surface, as shown in Fig. 25. The ripples passing under the right

Fig. 25. Diagram to illustrate the mechanism of vibrotaxis.

legs will have a smaller amplitude than those passing under the left legs, and thus there is a more intense stimulation of the right legs, followed by a greater degree of flexion of the femoral muscles. The consequence of this forced movement is that the extended left legs exert a turning effect relatively to the less extended right legs so that as the spider runs it is automatically directed towards the origin of the disturbance.

The distinction between kineses and taxes is well illustrated by considering movements determined by the force of gravity. An arachnid which habitually climbs upwards or downwards shows a discrimination between "up" and "down", which is a response to the direction of the stimulus. This is pure geotaxis. As it climbs up (or down) the force of gravity does not alter by an appreciable amount. Therefore the animal does not move more and more quickly as it approaches the top (or the bottom) of a tree trunk. If it did so, it would be exhibiting geokinesis. Because it does not, there is no geokinesis in arachnid behaviour.

The essential difference between the tropism theory and the taxis theory is a difference in interpretation, the former being mechanistic, the latter teleological. If a biologist can convince himself, in spite of the impossibility of obtaining direct evidence of the animal's subjective state, that the animal has a conscious appreciation of sensations, then the taxis theory will afford an acceptable interpretation of many facts. For instance, Kühn states that flies move towards the window if they are chased and that on the theory of tropisms such a sudden change in the sign of the reaction is inexplicable, as in any other such change when it appears as a result of suddenly approaching danger. As an innate reaction to "the simple sensation of danger", such behaviour may be intelligible.

The present position with regard to the interpretation of the behaviour of invertebrates like Arachnida may be summarized as follows. The mechanistic theories, such as the tropism theory, were valuable because they pointed the way to a great deal of experimental work, most if not all of which may be broken down into measurement of the reactions of the sense organs and the response of the muscles. It may be useful—perhaps it may be necessary—to know the facts about these actions, but their relation to any psychic events is obscure and experiment tells us nothing about these. The tactic view of behaviour may lead a biologist to ask whether a creature which, for example, lurks under stones does so for some conscious reason, and then force him to give a negative answer simply because the taxes which guided it have been recognized and named.

Mechanistic theories of behaviour carry us very little further than this. Despite the persuasiveness of the mechanists, an animal does not seem to be a mechanical device; it seems to be much more obviously a sentient being, trying, now in one way, now in another, to survive amid the perils of existence.

The position is no more than a logical consequence of a very simple dilemma. Either there is nothing in animals that corresponds to the origin of thoughts in man, or there is something in animals in which originate thoughts, or such homologues of thought as may occur in cerebral ganglia. The nature of these thoughts and of the mind in which they are developed is a fundamental problem of biopsychology, to which this chapter can do no more than serve as one path along which that problem may be approached.

Reflexes, kineses and taxes may well be looked upon as patterns of mechanical behaviour, evolved and retained in response to the environmental conditions in which Arachnida can survive. One is inclined to

think of Arachnida as having chosen a situation where the physical
conditions are nearly constant; they are to be found in ditches, under
stones, leaves and logs of wood, and in caves. This apparently ascribes
the selecting agency to the activity of the animal. But the boot is on the
other leg: the real selector is the environment. It is not drought or flood,
cold or warmth that disturbs the invertebrate, but the change from one
to the other. Should the conditions alter so as to overstep the various
thresholds, the animal is affected and must respond. Thus it is inexor-
ably moved on, as if by a merciless policeman, until it reaches a station
in which it can remain. Here it is a prisoner, until further change releases it.

This conception of an animal, devoid of behaviour in a constant
environment, is, unfortunately, contrary to ordinary ideas and ordinary
experience. But it is a fact that *ordinary* experience is seldom derived
from the study of animals in their normal surroundings. A naturalist
usually meets small animals just after he has disturbed them. He be-
labours the hedges and ploughs up the leaves or at least goes crashing
through their neighbourhood, his titanic feet shaking their world, his
shadow obscuring the sun, his scent lingering behind him. Only a few
observers realize how large a proportion of an animal's time is spent
doing nothing.

Here is the true place of the study of tropisms in animal behaviour.
Any particular topism can, of course, be understood only after experimental study under laboratory conditions, yet in nature animals exhibit
tropistic behaviour although no mechanistic biologist stands by to
measure their angles of deviation.

These mechanistic reactions are supplemented by activities, usually
of a greater apparent complexity, which are collectively and conven-
tionally described as instinctive. Instinctive acts, like reflexes, do not
have to be learnt and they are adaptive in the sense that they tend to the
preservation of the animal and of the race when they are normally
performed in normal circumstances, but when circumstances change it is
often found that the animal neither modifies nor reverses its instinctive
procedure, continuing its course unchecked with useless or even fatal
results. But, unlike reflexes, instinctive acts demand an intact nervous
system, and they are often directed towards a relatively distant goal.
Many of the deeds that animals perform on behalf of their young will
bear fruit only in the future, while reflexes are always concerned with
affairs of the moment. Further, instinctive actions are sometimes capable
of modification.

All these characteristics of instinctive actions are well illustrated by
the Arachnida. The spinning of the spider's web, the building of the
false scorpion's nest, the making of an egg-cocoon, the courtship of

scorpions, Solifugae, spiders and false scorpions and the elaborate behaviour involved in the life histories of mites and ticks are all instances of typically instinctive actions. As long as we are studying the Arachnida we are dealing with instinctive actions in which their cast-iron routine is well displayed. The well-known experiments of Fabre (1913) and the even more critical studies of Hingston (1928) have uniformly emphasized this fact, that instinctive behaviour may develop in an irreversible sequence and is usually quite unable to deal with any situation outside the ordinary. This is a description of the nature of instinct and not a criticism of the animal's ability, for the instinctive actions normally suffice for the animal's survival. The young spider spins its first web quite perfectly—and a year or so later it may spin its last. The last will be no better and no worse than the first, it will not be spun more quickly or in a better place. The spider has not profited from past experiences, and so has robbed us of all the evidence we might have had that the experiences were conscious ones. All its life it has just spun—instinctively, irresistibly, irrationally—for it has no consciousness; it has no mind.

Such instinctive behaviour is, in general, largely governed by internal physiological conditions, just as tropistic behaviour is governed by external physical conditions. The habits of the water wolf spider, *Pirata piraticus*, illustrate this particularly well. These spiders spin a tube-like retreat leading to the water's edge. When the female has laid her first cocoon she remains in the tube but neglects to repair it. It is not that she never spins, for at regular intervals she detaches the cocoon from her spinnerets, holds it in her third pair of legs and turns it about while adding more silk to the outside. She will then lay it aside for a short while and add a little silk to her home, but her instincts are those of a mother, not of a builder, and the tube soon becomes a wreck. By the time that the first cocoon has hatched and the young have dispersed, the ovaries will be maturing for a second family, and shortly before this cocoon is laid, the spider reverts to her maiden habit and spins a new and perfect home. This process of alternating neglect and replacement is repeated between the second and third cocoons, after which the summer is almost over and the winter torpor is approaching. When a number of these spiders are being kept at the same time, the influence of approaching and receding maternity upon their normal instincts is particularly striking.

Thus one important aspect of the nature of instinct begins to appear. To the older naturalists this presented such great difficulties that they were usually forced to be content with description and metaphor. Thus they spoke of instinct as "racial habit" or as "inherited memory" or as "lapsed intelligence", but these were mere verbalisms and no progress

results from sheltering behind words. Some advance has now been made
from that position, so that we now look upon instinctive actions as being
reflex acts, following one another in predetermined succession, coming
often under the influence of external circumstances, so that a symmetry of activity is produced, and often also under the influence of
internal conditions, so that new types of activity constantly appear.
This union of external and internal factors and the realization of the
nature of the latter is important. Instincts are something more than
reflexes, but this "something" is of a material nature, secreted by glands;
it acts as a chemical compound distributed by the blood. Thus it becomes at once subject to direct study, and mere difficulties of technique
are the obstacles to a full knowledge of the nature of every hormone, of
the way in which it is produced and the physicochemical nature of its
mode of action.

An exceptionally fine analysis of an instinctive action in this way has
been made by Peters (1931-3), in his study of the capture of insects by
the common spider *Araneus diadematus*. He concludes that a series of
stimuli, each followed by a characteristic reaction, produces the habit,
and he has broken down a process, which many have watched and
which appears to be a continuous operation, into the following steps:

STIMULUS RECEIVED	REFLEX RESPONSE
1. Vibration of threads of web	Movement to centre of disturbance
2. Struggling of prey	Long bite
3. Contact with prey	Wrapping or enshrouding
4. Contact with silk wrapping	Short bite
5. Chemical stimulus of short bite	Wrapped prey is carried off

An analysis, similar to this in every essential, has been made by
Homann (1928), who studied the approach of the jumping spider
Evarcha blancardi towards its victim. Here again the outward appearance
is one of a steady stalking, and here again the separate steps have been
recognized and isolated:

STIMULUS RECEIVED	REFLEX RESPONSE
1. Image in posterior lateral eye	Turning of spider's body towards object
2. Image in anterior lateral eye	Turning of spider's body continues
3. Images in both anterior lateral eyes	Spider begins to move forwards
4. Images in anterior median eyes	Spider begins to creep towards victim, or to court a female

Finally, an aspect of arachnid behaviour which appears as a characteristic feature in the lives of many is due to the rhythmical nature of individual activity. It has been said above that an arachnid spends a large proportion of its life waiting, and doing nothing; to this there must be added a recognition of the fact that idleness is often interrupted at regular intervals.

Rhythmic behaviour, which is by no means limited to the Arachnida, is of two kinds. There are exogenous rhythms, which are responses to regular recurring changes in the environment; and these are to be contrasted with endogenous rhythms, consequent upon internal changes within the organism. The former tend to disappear if the particular stimuli are removed; the latter persist even when the animal is kept under constant conditions, artificially maintained.

Probably the most familiar example of rhythmic behaviour among Arachnida is the simple observation that the spiders in our gardens are usually seen to be spinning their webs in the evenings, while there are other species that are equally accustomed to spin their webs just before the dawn. An example of precise observations of this kind of rhythmic behaviour is that described by Marples (1971) concerning the common species *Zygiella atrica*. The habit of this spider is to spend the day in shelter, though in communication with its web, and to take up its position in the centre of its web during the night.

Hourly observations of 127 webs gave the following result:

5.30 a.m.–10.00 p.m.	No spiders in the webs
10.00 p.m.–10.30 p.m.	Spiders beginning to occupy hub
10.30 p.m.–11.00 p.m.	All spiders occupying hubs
4.00 a.m.– 5.30 a.m.	Spiders spinning webs

Behaviour of this kind, clearly associated with the intensity of the light, is common enough among Arachnida of many orders. It is particularly obvious among Opiliones, as may be seen by anyone who keeps harvestmen in cages. Their rhythms were examined by Edgar and Yuan (1968), working on *Phalangium opilio* and seven species of Leiobunum, with the use of a kymograph. Their results showed that under all ordinary conditions 90% of their activity occurred between 6.00 p.m. and 6.00 a.m.

Fowler and Goodnight (1966), working with *Leiobunum longipes*, compared its activity with the secretion of 5-hydroxytryptamine. This compound is widespread in animals and plants, its presence can be readily shown and the amount measured.

The results obtained showed that activity began about two hours before sunset, rose to a maximum at about 2.00 a.m., and then rapidly declined to zero just before dawn. The content of 5-HT showed the same peak at 2.00 a.m. The distinctive feature, which agrees with that found in other animals, is that the occurrence, rather than the duration of activity, is constant.

Spiders and harvestmen are not exceptional among arachnids in this respect, for a similar rhythm is detectable in most, if not in all the other orders. As an example, one may quote the observation of Weygoldt (1971) that the Amblypygi in his laboratory showed three peaks of activity during the day, the first and most pronounced just after sunset, a second before dawn, and the third and weakest at about noon. Similarly, Cloudsley-Thompson (1973) has described the activity of the scorpion *Buthotus minax*, which rises to two maxima during each day, by reacting to the change from light to dark and, in daylight, to rising temperatures.

The activity of many animals is similarly regular, but the mechanism that controls them is often unknown. The rather abrupt changes that are to be observed suggest that a series of neurophysiological events must be involved. In any case the similarities between vertebrate and invertebrate behaviour, taking the form of response to the internal environment, is extremely interesting.

9
Zoogeography: Distribution

The distribution on the earth of any group of animals is the result of a combination of the place of origin, the means of travel and the nature of the obstacles encountered *en route*. The zoogeography of Arachnida is no less interesting than is that of any other class.

Early considerations suggest that four categories can be recognized.

(i) Arachnida which, though confined to the hot tropical or very warm sub-tropical belt are, however, more or less continuously represented therein—scorpions, theraphosid spiders, Amblypygi and Uropygi.

(ii) Arachnida which are also tropical or sub-tropical, but which are sporadically and discontinuously distributed therein—Palpigradi, Ricinulei, Schizomida and liphistiid spiders.

(iii) Arachnida which spread to the limits of the temperate zones—Opiliones, Pseudoscorpiones, Acari and most spiders.

(iv) Arachnida which do, in fact, spread into the polar regions—certain spiders, Opiliones and mites.

These may be taken in turn.

It is, in fact, very obvious that in the tropics the Arachnida are represented by some of their largest members. The scorpions, Solifugae and theraphosid spiders are groups which average more than 2 cm in body length, a great contrast to all the other orders, in which the body length averages less than a centimetre. This generalization is in striking contrast with that usually known as Allen's rule, which states that in any one group the polar members are larger than the tropical ones. This rule, however, was based on observations on warm-blooded mammals. The ratio area : volume is necessarily smaller for a large animal than for a small one, so that the former loses heat less rapidly than the latter. It is clear that for cold-blooded animals like Arachnida the rule cannot be accepted.

The most remarkable of the tropical Arachnida are those that are able to survive in the hostile conditions of the deserts that cover so much of the zone. Here the chief features of living must be the scarcity of food, the lack of water, the heat of the sun by day and the difference between day and night temperatures.

The large size mentioned above must be regarded as one form of adaptation to high temperatures. It is in itself a water-conserving mechanism, exploiting the low ratio between surface area and volume, and it is supplemented by the nature of the respiratory system. The book-lungs, which form the primitive respiratory organs of Arachnida, are most numerous in the scorpions and in the theraphosid spiders: they can function efficiently in connection with a less elaborate circulatory system than that which must exist in an arachnid with a localized respiratory system.

The large desert Solifugae represent an alternative arrangement. They have no book-lungs, but only tracheae. Their tracheal system is, however, more elaborate than that found in any other order, and thus permits the great activity and high speeds that are so characteristic of this order. A tracheal system of respiration is, in fact, an advance on the book-lung, and has been very successful in other orders.

The sporadically distributed species of the first three orders mentioned in category (ii) contain the most strongly cryptozoic, photophobic, stereotropic and hygrophilic of all Arachnida. Most of them are slight in build, slender in dimensions, delicate in constitution; and they are among the most typical of all Arachnida. They contrast most emphatically with the great multitude of temperate zone species, which include, among others, the web spiders that live exposed lives in the open.

The very large third group, inhabiting both north and south temperate zones and reaching its limits towards the polar regions, illustrates very clearly the relationship between physical conditions, body size and the respiratory system.

Book-lungs are in the majority limited to a single pair served by a more complicated arterial system, but very often they are supplemented by tracheae. This is true of the majority of spiders, which also present differences in the position of the spiracular opening. There can be little doubt that tracheae help to economize internal moisture, and that in small spiders this function is improved by a thicker sclerotization of the exoskeleton.

The orders Opiliones and Pseudoscorpiones show tracheal respiration only, and in the most active genera of the former there are supplementary tracheae with spiracles in the legs.

The fourth category, the polar Arachnida, is also well characterized. There is a sharp distinction between the Arctic and Antarctic, due to the isolation of the Antarctic continent and the tempestuous nature of the oceans that surround it, an isolation that affects the character of other groups in the polar faunas.

The first record of an Antarctic arachnid was the discovery of the mite *Penthaleus belli* under moss at Cape Adare during the Southern Cross expedition of 1899–1900. During Shackelton's Nimrod expedition (1907–09) several species of mites were found to be abundant in Coast Lake, near Cape Royds on Ross Island. Exoskeletons of others were found in other lakes, but none were seen alive. A mite, however, hatched among vegetable matter brought from Deep Lake after the return of the expedition to England. These Ross Island mites were never named in the scientific reports of the expedition, but they are of interest as being among the first Arachnida to be recorded from the far south. Since then, mites and accompanying spring-tails have been found at stations from sea-level to heights of 1,830 m in latitude 77°South, where the temperature may fall in the winter to −65°C, and as far inland as the 86th parallel. Mites appear to be more resistant than insects and so able to survive by living under stones where the temperature remains high enough to allow metabolism to continue.

No spiders or members of any other order have been found on the Antarctic continent, and the most southerly spider is most probably either *Rubrius subfasciatus* from Cape Horn or *Notiomaso australis* from South Georgia in 55°S.

Within the Arctic Circle are lands of comparative fertility, supporting the Esquimaux and the Samoyeds, with their herds, and producing flowering plants. The 80th parallel of north latitude passes through the island of Spitzbergen, and a very considerable number of spiders and at least one harvestman have now been recorded from this island and elsewhere, chiefly by the pioneering researches of Braendegaarde. The farthest north land is the northern coast of Greenland, which reaches latitude 83°N, only 644 km from the Pole. This region was explored in 1917 by the Fourth Thule expedition led by K. Rasmussen, and their report stated that "Spiders and Earth-mites also support life in these high latitudes". At least 50 species of spiders have been found in Greenland, and 15 species are known from the island of Spitzbergen.

In comparison with these results it is interesting to recall that in 1924 spiders of the family Salticidae were found at a height of 6,700 m on Mount Everest, and were then described as holding the proud position of being the highest permanent inhabitants of the world. They must, however, be supposed to share this honour with some insects, probably spring-tails, which, though undetected at the time, provide them with food.

Two other aspects of distribution remain to be considered, the unexpected and the cosmic.

Man is the only animal that constructs devices or machines that enable him to travel. Wheeled vehicles, ships and aeroplanes have all helped him to carry himself and his impedimenta wherever he has wished to go. And very often Arachnida have accompanied him, unawares.

Many of these excursions have been trivial and of no significance because the stowaways have been unable to survive in their new surroundings: familiar examples are the so-called "banana spiders", which periodically arrive at British ports in bunches of bananas.

To these may be contrasted the fair number of spiders recognized as members of the British fauna, which have reached this country in recent years. They include *Ostearius melanopygius*, *Hasarius adansoni*, *Achaearanea tepidariorum* and *Argiope bruennichi*. Members of other orders are known, if more rarely, to have made similar journeys, successfully.

The surprising report in 1914 by Berland of a species of Eukoenenia apparently acclimatized to life in the Natural History Museum of Paris was followed in 1933 by the finding of *Eukoenenia mirabilis* on the lower slopes of Mount Osmond, Adelaide, and quite recently Marples has collected others in Samoa. Accompanying them in Samoa there were some Schizomida, while Cooke and Shadab (1973) have found immigrant Uropygi in Gambia, and Weygoldt (1972) has reported Amblypygi that have been living in the Dodecanese Islands of Rhodos and Kos.

Finally there may be mentioned the support given by Arachnida to Wegener's theory of Continental Drift. Legendre (1968) has written a fascinating essay on Gondwanaland and its problems, in which he refers to a number of significant examples.

Forster (1949) reported the close affinity between certain New Zealand Cyphophthalmi and members of the same order found in South Africa and South America.

The order Ricinulei is known from the Amazon basin, the caves of Mexico and West Africa.

Among theraphosid spiders, Zapfe (1961) pointed out the occurrence of the sub-family Miginae in Chile, Australia and New Zealand. Legendre himself instances the family Archaeidae, discovered in Baltic amber before it was found living, whose type genus Archaea is known from South Africa, Madagascar, New Zealand and Australia.

As striking as any of these is the distribution of the four remarkable genera of spiders that catch their prey by means of a ball of silk at the end of a swinging thread. These "bolas spiders" belong to the genera Dichrosticus in Queensland, Mastophora and Agatostichus in America and Cladomelea in South Africa.

10

Ecology: Migration and Dispersal

The dispersal of animals of any group throughout the country of its occupation depends ultimately on two factors: the means of travel adopted by or available to the individual, and the physical conditions that it encounters during its travels. The first, controlling the distance that it can go, determines the area that can be colonized; the second determines its survival at the end of its journey.

Few Arachnida are great travellers. There are many species of scorpions that scarcely trouble even to move away from one another and have been somewhat loosely described as living in colonies. But there is no social organization in these gatherings: scorpions are naturally lazy creatures, and if food is available they will not travel far. Hunger is often one of the chief stimuli to activity.

Among spiders one method of migration stands out, the well known gossamer flight. A spider about to migrate shows what seems to be an irresistible desire to climb; it scales the nearest vertical object, be it gatepost, railing or tree, with a persistence that has all the outward appearance of determination and purpose. When it reaches the summit it turns round so that it is facing the direction of the wind, and, raising its abdomen, secretes a droplet of silk from its spinnerets. The wind draws this out into a long, buoyant streamer, and soon the spider releases its hold and flies away.

There is no cause for surprise at the interest that this method of dispersal has aroused for so long, for it is without parallel among other animals. Dispersal by drifting is common enough and is shown by all the multitude of marine things that compose the plankton of the seas, as well as by those that similarly float through the air. This aerial plankton has been found to include spiders, plentifully up to heights of 61 m and occasionally up to 3,050 m. But these planktonic creatures are drifting, like seeds and spores, only because they are so small and light; none of them has produced an extended current-catcher like the spider's long thread. This thread is comparable to a parachute or glider, capable of transporting for hundreds of kilometres an animal which, without it, would scarcely be wafted as many centimetres.

In the spring and autumn, seasons in which large numbers of spiders are hatched, gossamer is more than usually prevalent and carries chiefly young specimens; in the summer the aeronauts include many half-grown and mature examples of the smaller species. The earliest opinion was that one species, "the gossamer spider", was responsible for all these threads, but this is not so. Emerton in 1918 found 69 American species in the air, Bristowe in 1930 listed 30 British ones belonging to six families, and Bishop in 1945 identified 25 species in six families from heights between 6·1 and 1,524 m. It is believed today that nearly all the families contain species which undertake aerial travel at some stage in their lives, though it is clear that this cannot be expected of the largest, which are too heavy to fly.

Even so, it cannot be maintained that the "ballooning habit" is common to all, or even to the majority of spiders. Take Emerton's total of 69, Bristowe's total of 30, and Bishop's total of 25; add them together and then, not to seem parsimonious, multiply the sum by ten. This makes a generous allowance for more observers in different countries, but the product is only 1,240, a very small fraction of the number of species of spiders known to exist.

A different method of dispersal is shown by some of the smallest mites that are parasitic on fruit trees. These migrate in the summer, either by grasping the legs of passing insects, or by standing erect and leaping vertically if they feel a puff of air. The leap may then help them to be blown some distance. The common British mite *Belaustium nemorum* is often found holding the legs of Tipulidae; and several species of uropod mites attach themselves by a thread of their own excrement to the bodies of beetles.

A more peculiar method of travel is found among false scorpions, and is known as phoresy. False scorpions are essentially lovers of darkness, and yet on occasions they are to be seen clinging to the legs of flies, harvestmen and other creatures, and thus being carried about at no effort to themselves. They grasp the fly's leg in their chelicerae but they are not parasites, for they neither injure nor feed on the fly. Twenty-five or more species belonging to about 20 genera have been found in these circumstances. The habit is not equally prevalent at all times of the year, but during the short time that it appears it is reasonably frequent. Thus Vachon (1932) recorded the finding of 78 specimens on the legs of 57 harvestmen in one week at the end of August. All belonged to the same species, and all but one were mature females. The highest number found on the legs of any one harvestman was eight.

The maturity of the travellers shows that the habit of phoresy has no relation to the dispersal of a brood of youngsters. Vachon found that all his captures were either recently fertilized or alternatively that they had just left the brood-chamber in which they had reared and fed their family. In either case they were in urgent need of food, and he concluded that hunger had made them seize the legs of flies or harvestmen, and that as a result they reached areas where food might be plentiful.

An alternative view, which has the support of the authority of Muchmore (1971) is that the false scorpion grasps the fly's or harvestman's leg in its search for food, and that the bearer moves on before the passenger lets go. The consequent dispersal is therefore only accidental.

A word should be added concerning plain, unspecialized wanderings. The impression that may be given by the last few paragraphs, or by the casual statement that Arachnida can escape from their kin by merely running away and seeking fresh fields, is one that can easily be exaggerated. A reader may think of such an animal as one that has no home, and which, wandering continuously, may find itself many miles from its birthplace. This does not, in fact, happen. Entomologists who have painted coloured spots on ordinary house-flies, so that they can be recognized if they are caught again, find that flights of a kilometre or so are the exception.

Even the wolf spiders that one sees in a meadow are likely to have been born in it and are as likely to die in it: the individual, in fact, lives normally in a circumscribed area over which it hunts and where it is too busily occupied in daily life to have the opportunity for foreign travel. This is conspicuous when wolf spiders are kept under observation in the laboratory. In a large round pneumatic trough a solitary spider has an area of about 2,600 cm² of sand and gravel, where it energetically hunts and catches any insects that may be offered to it. And somewhere in this field it has found a spot, under a stone or between two stones, which, though to our eyes no different from other spots, has attracted it and has become its "home". Here it usually rests when not on the prowl, and here it returns after its forays, displaying a homing instinct or mnemotaxis. This description of the act of homing should be compared with the account of the function of the lyriform organs, given in Chapter 3.

Migration in such ways as these is so largely dependent on chance that it is not inevitably followed by survival in the new surroundings. The migrant must be adapted to its new circumstances, and the methodical study of the relations between the organism and all the features of its environment has grown into the established science of Ecology. This may be said to present two sides, distinguished as autecology and synecology.

Autecology concentrates attention on the lives of the separate species and on their reactions to physical conditions. These include the prevalent temperature and its range of variation, the humidity of the atmosphere, the intensity of the light, the velocity of the winds and the chemical constitution of the soil; all of which are together described as the edaphic factors. Synecology is concerned more with the community or population than with the individual. Thus it takes account of predators and parasites, of food chains, and of changes in the density of population during the year. These are known collectively as the biotic factors. Clearly it is less elementary than autecology, a characteristic of which is that the effects of physical conditions can often be most precisely determined by experimental study in the laboratory.

The synecologist faces a more complex situation, which is most readily approached by a convenient division of the biosphere into three layers.

(i) The Gound Layer: the moss and litter layer, up to about 15 cm above the surface.

(ii) The Field Layer: the plant and shrub layer, up to about 1 m above the surface.

(iii) The Canopy: the tree layer of trunk and branches, as well as walls and the outsides of buildings.

The lowest of these layers is the home of many more individual animals than either of the others, for in there are to be found not only those that are actually underground, but also the vast assembly known collectively as the Cryptozoa. These are the animals that lead lives in conditions that are continuously dark and damp, and in which conditions are less changeable than elsewhere. This constancy is favourable to most Arachnida, and in fact Lawrence (1953) estimated that about half the cryptozoa belong to this class. Every field naturalist knows that the sifting of leaf-litter yields spiders, mites, false scorpions and harvestmen, while in the right parts of the world there may also be Palpigradi, Schizomida and Ricinulei.

A conspicuous feature of all cryptozoa is the primitive nature of many of their number, relating them to their distant ancestors, who long ago crept from the primaeval ocean. They found in the cryptosphere an environment in which survival was possible for small animals with thin exoskeletons, and tactile and chemotactic sense organs that made blindness of small consequence. Life in the cryptosphere was serene and there was little temptation to leave it and to evolve into more specialized forms. From this arises the statement that whereas the vertebrate palaeontologist looks for fossils the invertebrate palaeontologist looks for cryptozoa. The cryptozoic lives of the Palpigradi and Schizomida are examples of this.

On the other hand, spiders are among the most highly specialized of the Arachnida, and among the cryptozoic spiders the families Linyphiidae and Erigonidae are prominent. They illustrate the advantages of small size. In addition there are many spiders that may be found in the soil itself. Estimates of their number have ranged from 11,000 to 600,000 spiders per acre. And there are some spiders which, though normally living on the surface, have adopted a habit of burying themselves and thus securing protection. Reiskind (1965) has described the steps by which a species of Sicarius digs a hole in sand with its forelegs, enters, and throws sand over itself until it is covered.

An environment that may be considered as an extension or modification of the cryptosphere is that of the cave. A cave offers the same conditions of darkness and humidity, and the further a cave may extend from its entrance, the less are these conditions likely to change. The occupants of caves include a proportion of Arachnida, which share with members of other classes the features that may be attributed to the environment.

The high relative humidity, which often reaches 100%, results in a slower metabolic rate. Since oxidation helps to produce melanin, cave-dwelling fauna are often pale or colourless. For the same reason, they are often sluggish in movement and less aggressive in behaviour; and this is associated with the reduction in the proportion of chitin in the exoskeleton and also with an increased length of body and legs. The blindness, which is common, is sometimes attributed to the loss of eyes owing to disuse, or alternatively to the fact that organisms, eyeless from some other cause, are at no consequent disadvantage in a cave.

Probably the most interesting relation between caves and Arachnida is shown by the Ricinulei. Long regarded as the rarest representatives of their class, these animals have been found by Mitchell (1970) to be living in their thousands in the caves of Mexico. In the words of one arachnologist, "They were so numerous that it was hard to avoid treading on them".

To this there should be added the report by Mitchell (1968) of the discovery of three eyeless scorpions, also in Mexican caves. For these, the first blind scorpions recorded, a new genus, Typhlochactas, and a new sub-family were necessary. The species are described as being some 15 mm long, smooth and weakly sclerotized, the exoskeleton so translucent that the alimentary canal could be seen within. The occurrence of such unusual scorpions, at the considerable height of about 1,400 m poses interesting problems in the evolution of the order.

The Arachnida that live in the upper, field and canopy layers, cannot secure the same security from the nature of their immediate surroundings. They are therefore often found to exhibit such adaptations as protective colouring and mimicry, which reduce their risks of falling victims to predators. Recent study of the Arachnida of this region has revealed two characteristic and unexpected features in the lives of its occupants.

The first of these is a migration from the ground to the trunks and lowest branches of the trees in the spring, to be followed by a return to the lower levels in the autumn. Here there is an alternative. The animal may lay eggs in the ground and die, or it may hibernate underground and awake to lay eggs in the spring. This is the normal timetable for a number of spiders, and probably for some of the harvestmen and false scorpions as well, which increase in numbers at ground level as the autumn sets in.

A more complex relation has been found to exist between two species of spiders of the genus Clubiona, and has been recorded by Duffey (1969). *Clubiona brevipes* can be shaken from the foliage of trees in the late spring and summer; in autumn it moves down to the trunk and hibernates in crevices in the bark. Complementary to this, *Clubiona compta* spends spring and summer on the bark and comes down to the ground in the autumn. These rhythms tend to prevent competition between the two species.

The second and more surprising feature of environmental preference is the occurrence of two or more species in two apparently different niches, to which, however, they are limited. The spider *Tibellus maritimus* first drew attention to this when it was found to occur only in grass roots on sand dunes or in the much moister conditions of marshland. This peculiar contrast it was later found to share with the three species *Hypomma bituberculata*, *Synageles venator* and *Clubiona juvenis*. These species, confined as they are to two such differing environments, were described by Duffey (1968) as being diplostenoecious.

Diplostenoecism deserves more investigation. It points to unknown factors in environmental choice, and can scarcely be limited to spiders. Other orders of Arachnida and other classes of Arthropoda are almost certain to contain further examples.

There is much ecological work to be done in the future on the adaptations of all orders of Arachnida to life in the larger regions of special character. Environmental peculiarities exist not only in ditches, caves and ants' nests, but may cover much broader areas, presenting greater problems both to the occupying Arachnida and to the investigating arachnologist. Among such are mountains, deserts, sand dunes, oceanic islands, the littoral region, houses, towns and cities, river banks and so on.

From all these there has come a not inconsiderable amount of information about the spiders that have been found in them, with much fascinating description of many remarkable ways in which spiders have conquered the environment. The proper place for a survey of this ecological knowledge is in some comprehensive account of the biology of spiders; one single order cannot claim the necessary space here. When one looks, for example, at the existing work of Lawrence, Vachon, and Cloudsley-Thompson on the scorpions of the African deserts, and reads of their nocturnal activities and their daily search for protection, one begins to realize how much is to be discovered about the living conditions of the other widely distributed orders. Here lies the greatest challenge, or the greatest opportunity for all ecologically-minded arachnologists.

Dondale and Hegdekar (1975) have admirably expounded the task that faces the arachnological ecologists.

The survival of any arachnid in any environment is possible only if it can find protection from adverse physical and climatic conditions, protection from enemies, a sufficient quantity of food and the opportunity to reproduce its species. Satisfaction in these respects depends on mutations and natural selection, and constitutes the first principle of ecology.

A second principle involves the response of a species to changes in any of these circumstances. If the reactions are favourable they may bring about an abundance of the species in the area; if, on the contrary, they are less fortunate the result is a falling population and perhaps ultimate extinction.

A vital factor in the survival of any species concerns its powers of dispersal, closely linked with which is the chance that the migrants may find their new area friendly or hostile. It may be that all the arrivals quickly perish; but often it is to be found that they occupy their new habitats during certain seasons of the year only, or until the environment changes adversely. Lastly, there may be some arrivals that succeed in becoming permanent residents.

The number of these established species may well increase, and the whole consortium become more and more complicated as more species find their way to vacant niches, to which they are already pre-adapted. This leads to the overriding problems.

The ecologist cannot understand or satisfactorily explain the whole of this multiple complex until he has determined the life history of each inhabiting species. Not until he has this knowledge can he realize what factors determine the adaptation of a population to a given area, whether the different constituent species are continuously or sporadically present and are widespread or circumscribed within it. This is no mean task for the future of arachnology.

11

Phylogeny: Evolution

As has been foreshadowed in Chapter 2, the evolutionary problems which the Arachnida pose are unusual in character and difficult to solve, for the class is one of which the origin is uncertain and within which the succeeding, progressive, changes are obscured. With as many as 17 reasonably well defined and recognizably different orders, an arrangement in a series of ascending steps from the earliest and most primitive to the latest and most elaborate ought not to be impossible, yet, in fact, it is.

The chief difficulty lies in the nature of the palaeontological record, which does not present us with a series of fossils, dated, at least in relation to each other, by the strata in which they have been found.

An arachnologist's first concern is with the Silurian period. Its predecessor, the Ordovician, had been marked by a great increase in the number of living things and simultaneously in the appearance of new species. This was particularly true of the marine invertebrates, the land being occupied only by plants and probably no insects. The Silurian rocks were mostly deposited in the littoral and shallow waters of the seas, and they show us how the evolution of the animal kingdom continued with undiminished vigour. The seas were populated by Eurypterida (Fig. 26), which now began to reach the great size for which they are notorious; Xiphosura such as Hemiaspis abounded, and the Ostracoderms foreshadowed the arrival of the vertebrates.

It was during this period that the scorpions appear to have left the water and established their claim to have been the animals that began the colonization of the land. They are usually regarded as descendants from an eurypterid stock, and perhaps they were driven to their momentous experiment by the rigours of life in the Silurian seas. Overcrowding invariably leads to a state of pollution, such as we know only too well today, and in general provokes a reaction which takes the form of emigration. But whatever the stimulus and whatever the conditions of survival, the order of scorpions has been consistently separated from the

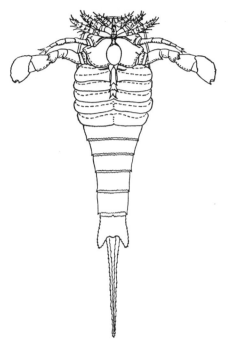

FIG. 26. *Eurypterus fisheri*. After Holm.

other orders since the time of Ray Lankester's classification in 1905.

These pioneer scorpions may have been able to survive the traumatic change of environment because of their chitinous exoskeleton, which retarded the loss of water from the body and prevented a fatal desiccation which even today is a constant danger to all land Arthropoda. The necessary alteration of the respiratory system was relatively uncomplicated, for book-gills which developed a covering would thereby be converted into book-lungs. The close histological resemblance between the gills of king crabs and the lungs of scorpions supports this idea. Two points, however, remain to be considered.

The first is the revolutionary hypothesis, put forward by Versluys and Demoll (1922), that the terrestrial scorpions were the first of the Chelicerata and were the ancestors of the aquatic Eurypterida and their allies.

The second point is the undeniably close resemblance between Palaeophonus, the Silurian scorpion, and its modern counterparts. True there is a peculiar difference in the ends of the tarsi, but apart from this the earliest scorpions have the large well-formed chelicerae, the segmented opisthosoma in which mesosoma and metasoma are clearly distinguished, the same mysterious pectines and the same terminal sting. These are not characteristics of a primitive animal, so that one is forced to imagine a hypothetical ancestor, a truly primitive scorpion from which the Silurian forms were descended. This imaginary proto-scorpion can more easily be imagined to be the ancestor of the Euryp-terida than can any known extinct scorpion.

Nevertheless, the hypothesis is not one that calls for immediate recog-nition. Migration from the land to the water is not to be readily accepted, nor does any Eurypterid suggest a scorpion-like ancestry. In the absence of any fossil scorpion earlier than the Silurian the specu-lation cannot usefully be pushed further.

There is a frustrating lack of evidence from the ensuing Devonian epoch, with very few arachnid fossils. This period was the first to have left evidence of great changes in the environment. The seas continued to be overcrowded; Pterygotus was prominent, and fishes were so numerous that the time is often known as the Age of Fishes. On land, invertebrate life developed: there were Ephemeroptera with a wing span of 13 cm, and an intelligent observer, following the course of foretelling the future by recollection of the past, would have prophesied a mass invasion of the land.

It came in the Carboniferous, with nothing less than an arachnid explosion. A warm, moist atmosphere favoured a territorial existence; plant life was luxuriant, insects and myriapods were accompanied by nearly all the orders of Arachnida that we recognize today. For many marine animals life in the sea was becoming less and less tolerable, and for the Eurypterida at any rate it was about to pass the acceptable limit. They were destined for extinction, and did, in fact, become extinct in the succeeding Permian epoch. The temptation to go ashore in search of better conditions must have been great; and, as Goethe has said, "Animals are always attempting the impossible, and often achieving it". In the end their achievements resulted in the creation of some 16 orders, the geological distribution of which is displayed in the follow-ing, familiar table.

Such a fragmentary record does not give as much help as might be expected. It shows that most of the orders were in existence during the Carboniferous era, and that they largely evolved along different lines, as a result of which five of them disappeared and 12 survived. The most significant fact that emerges from this view of the palaeontology of Arachnida is the conviction that they provide a good example of that general parallelism in evolutionary progress which is such a conspicuous feature of evolution as a whole.

The idea of the evolution of early scorpions from contemporary

	PRIMARY						SECONDARY			TERTIARY					RECENT
	Cambrian	Ordovician	Silurian	Devonian	Carboniferous	Permian	Triassic	Jurassic	Cretaceous	Eocene	Oligocene	Miocene	Pliocene	Pleistocene	
Eurypterida	x	x	x	x	x	x									
Xiphosura	x	x	x	x	x	x	x	x	x	x	x	x	x	x	x
Scorpiones			x	x	x		x				x				x
Palpigradi							x								x
Uropygi				x											x
Schizomida													x		x
Amblypygi				x											x
Araneae				x	x			x	x	x	x	x			x
Kustarachnae					x										
Trigonotarbi				x	x										
Anthracomarti					x	x									
Haptopoda					x										
Architarbi					x										
Opiliones					x				x		x				x
Cyphophthalmi															x
Acari			x								x		x		x
Ricinulei					x										x
Pseudoscorpiones											x				x
Solifugae					x										x

Eurypterida seems to have been accepted so whole-heartedly that further investigation has been neglected; yet the taxonomic isolation of the order Scorpiones should have indicated the need for some interpretation. The scorpions cannot be regarded as the ancestors of the other orders, and this suggests that during the late Devonian or early Carboniferous many other occupants of the increasingly unpleasant sea must have followed the example of their predecessors and repeated their historic landing. There is no reason to suppose that all these were of the same species, or even of the same family; more probably they were of different types, and the terrestrial Arachnida that evolved from them were different in consequence. We call them different orders.

Fossil scorpions seem to arrange themselves in nine super-families, only one of which can claim relationship with the living scorpions of the present. One may say, in broad outline, that ancestral scorpions of nine kinds made the experiment and that only one succeeded. Let the same idea be applied to the later invasion, and it may be said that 16 explorers set foot ashore and 11 of them founded permanent orders. Survival rate had risen from 11% to 69%.

One wonders why.

An obvious assumption is that the land was in a condition more favourable to sustain life than it had been at the earlier period, and that, correspondingly, the new arrivals were better fitted to exploit the improvements. And how better fitted? If we are right in guessing that they had come not from the depths of the ocean, but from the shallower or even from the littoral zone, they would already have become accustomed to exposure during low tides, when the use of atmospheric oxygen was essential. This is most likely to be possible for a small animal, seeking shelter under rocks, stones or seaweeds. If the relatively large surface of a small animal can be kept moist, the moisture may dissolve enough oxygen to avert catastrophe.

The danger to the pioneers came, therefore, not so much from the difficulty of respiration as from the risk of desiccation, as formidable a threat then as it is today. Arachnida and other Arthropoda can avoid its worst effects because of a layer of wax in the cuticle. In marine organisms such a layer may have controlled the passage of water by osmosis in and out of the body. This would have been particularly valuable to esturine species, exposed to recurrent changes in the salinity of the surrounding water. This might be claimed as an example of pre-selection for survival in a new environment, where the immigrants would swell the ranks of the animals, still in the majority, that are known collectively as the cryptozoa.

Nor should the possibility be neglected that some of the survivors did not leave the sea voluntarily, if one may use the term, but were carried ashore by tidal waves or in spray during an onshore gale. Cataclasms of this kind have been adduced by other zoologists to explain some of the peculiarities of evolution and distribution. Small organisms, thus ill-treated, might survive the experience, while many others must have failed to do so. It appears that 16 kinds succeeded.

These products of speculation and imagination may well be followed by, and indeed contrasted with, more orthodox attempts to trace the evolutionary steps by which the orders have become as distinguishable as they are today. The Arachnida which combine the greatest number of primitive features are the Palpigradi; it is also undoubted that the greatest number of specialized features is nine or ten, the total found in the Ricinulei. The many structural differences between these two extremes guide us in our search, and the studies of Petrunkevitch (1949) suggested one way of attempting this reconstruction of the past.

There is no arachnid in which all the prosomatic tergites are separate, and in nearly all orders the prosoma is covered by a uniform carapace. However, in the three orders Schizomida, Solifugae and Palpigradi there is a large propeltidium in front, with smaller plates, the mesopeltidium and metapeltidium, behind it. This suggests that fusion of the tergites has proceeded backwards from the head.

On the ventral surface the sternites have partly fused in some orders to form a large sternum, or have disappeared. These changes are associated with the movement of the mouth and the approach of the coxae towards the centre. There is always a labium or lower lip, the sternite of the second somite, and in the Schizomida, Uropygi and Palpigradi there are persistent sternites behind it. The backward movement of the mouth to a position between the pedipalpi, which can be clearly witnessed during individual development, is common to all Arachnida. In the extinct orders Haptopoda and Anthracomarti, as in the living king crabs, it is between the coxae of the first pair of legs. The coxae of the pedipalpi and legs take shares, in varying degrees, in the mastication of the prey, so that their approach to each other is not surprising, and in Opiliones and Solifugae no remnant of sternites is usually visible. The last, fundamentally the sixth, sternite sometimes persists between the fourth coxae, a position which, in Solifugae, is occupied by the first sternite of the opisthosoma.

In the opisthosoma changes have occurred in the first somite and some of the posterior somites.

The first tergite may be lost, as in recent scorpions, or fused with the carapace, as in the Opiliones. In the orders grouped as the Caulogastra it is constricted or reduced in circumference, to form the pedicel, seen in its primitive form in the Palpigradi.

The last three somites show different changes. They may be rather similarly constricted to form the narrow region known as the pygidium, to be seen in Uropygi, Schizomida and Ricinulei; but these regions are not strictly homologous, for in the Uropygi the pygidium represents somites 10, 11 and 12, in the Schizomida 9, 10 and 11, and in the Ricinulei 7, 8 and 9, or, perhaps, 8, 9 and 10. The last stage in the process is seen in spiders, where the anal tubercle represents from one to three persistent tergites, their sternites having vanished. In spiders other than Liphistiidae only five opisthosomatic somites remain, and in mites there is often no trace either of segmentation or of anal tubercle.

This reduction of the opisthosoma has been carried to its extreme limit in the Pycnogonida, a characteristic which well defines the relationship and the difference between the Pycnogonida and the Arachnida.

These evolutionary changes in the construction of the arachnid body may therefore be summarized by saying that in the prosoma there has been a gradual fusion of the tergites proceeding from before backwards, and in the opisthosoma a loss or fusion of somites proceeding from behind forward. It may be added that these changes appear to have taken place independently, and that they have tended towards a simplification of the primitive 18-somite pattern. At the same time specializations have appeared, making for greater distinctions between the different orders. Many of these are seen to be associated with environment and habit. As might be expected they are fewer in the orders with persisting segmentation, the scorpions, Uropygi and Amblypygi, and more numerous in those with the contrasting type of body, the spiders, mites and Opiliones.

Thus it seems possible to make reasonable suggestions as to the types of change that evolution has brought about, but impossible to put them into an acceptable chronological order, to show which came first and which followed. The only hints that we get in this are the contrasting facts that changes found to be common to several orders are presumably the older, and those limited to fewer orders are to be taken as more recent. However, the kinds of change provide some evidence as to the related problem of the origin of Chelicerata and of Arachnida, and an attempt may be made to summarize the whole story. It must, however, be made with a prefatory note of caution.

First there may have arisen among the ancestral Arthropoda an animal which may be described as a proto-chelicerate. It differed from its contemporaries in three ways: its mouth, originally terminal as in Annelida today, had moved to the ventral surface; the opisthosoma had become limited to 12 somites, each with its separate tergite and sternite; and thirdly there was a reduction of the middle portion or deutocerebrum of the cerebral ganglion, so that the "brain" appeared to have a reduced capacity.

These earliest Chelicerata were marine organisms, and from them as a starting point evolution seems to have proceeded in four directions. In one of these an extraordinary reduction of the opisthosoma took place, leaving little more than a plain sac, with few functions beyond a temporary retention of the contents of the hind-gut. These became the Pycnogonida.

Some Chelicerata developed in a less spectacular manner with changes in the shape of the body and a modification of the last pair of legs into paddles, useful for swimming. There was also a doublure or folding down of the fore-edge of the carapace, which had the effect of pushing the mouth back until it lay between the first pair of legs. These animals were conspicuously successful; they grew to an immense size, and their numerous fossil remains are known to us today as Eurypterida.

An allied group never took so determinedly to swimming, but crawled about on the mud and sand, helped by a telson in the form of a spine. They survived to remain with us as a small group of Merostomata known as the king-crabs.

The really dramatic step was taken by some littoral Eurypterida which began to leave the shallow water and to crawl on to the land.

The consequence was the evolution of the earliest Arachnida, a terrestrial branch of Chelicerata facing all the problems of life in dry air. Their conspicuous need was for a new method of respiration, and it seems to be very probable that if the book-gills of their ancestors had not been pre-adapted for conversion into book-lungs, Arachnida as we know them would never have become a possibility.

When once this step was taken, other changes followed. The prosomatic tergites fused into a more efficient carapace, strengthening the prosoma for its task of supporting eight legs. Segmentation remained, however, in the opisthosoma and has disappeared only in the highest groups. The posterior appendages vanished and the genital orifice stabilized its position on the second opisthosomatic somite (Bristowe, 1958).

Later division into the orders now recognized was determined by the production of a pedicel and by various changes in the mouth parts.

A view of arachnid evolution, wholly different from any of the above, has been given, at length, by Firstman (1973). He investigated the development and relations of the arterial system and the endosternite in many representative genera, and came to the conclusion that a relationship between these two systems is a primitive feature of the Chelicerata. From this there followed an emphasis on the contrast between the orders of Arachnida with book-lungs and those with tracheae only.

He concluded with a scheme involving neoteny, followed by adaptive radiation, which may be summarized thus:

(i) The ancestral scorpion was a neotenic Eurypterid, as is shown by the development of the lungs, limbs, lateral eyes, endosternite and arteries.

(ii) From this arose by adaptive radiation (a) modern scorpions (b) other pulmonate Arachnida.

(iii) Apulmonate Arachnida were neotenic scorpions, as is shown by the cephalothoracic tergites and sternites, limbs, respiratory organs, endosternite and abdominal tagmata, and they divide into (a) Palpigradi (b) other apulmonates.

Firstman's views are original and, backed as they are by much morphological analysis, deserve careful appraisal.

In conclusion, an attempt may be made to bring together the different ideas that have been expressed by students of arachnid phylogeny, and to form them into a single picture.

The first point that emerges clearly enough is the separation of the scorpions from the other orders. First on the scene, they have from the start preserved a form that has remained almost unchanged throughout the ages. In this consistency they recall the tortoises, a reptilian order with an equally unchanging equipment of genes.

For the rest, the outline of the story seems to read as follows.

Some newly-emerged immigrants were small enough to enable them to absorb enough oxygen for survival if their bodies remained wet. This occurred if they assumed a strictly cryptozoic mode of life, establishing themselves in constantly dark and damp surroundings. Such were the Palpigradi, Schizomida, Cyphophthalmi and Pseudoscorpiones (or the earlier, ancestral forms of these orders).

The appearance of tracheae not only facilitated the business of respiration, it was evidence of spontaneous or fortuitous changes in the gene pool of the pioneers. There followed a tendency to increase in size, which made possible, or perhaps necessary, an escape from the cryptosphere, and adaptive radiation into habitats of more variable natures. While the Palpigradi have, like the Scorpiones, retained a large proportion of primitive characteristics, the ancestral Schizomida gave rise to the early Uropygi, the Cyphophthalmi to the first Opiliones, and the Pseudoscorpiones to the Solifugae.

The next step was the abandonment of the spermatophore as a device for the impregnation of the females. In the flagellum-bearing group there came the loss of the flagellum, converting "Uropygi" into "Amblypygi", and finally, in the Araneae, the use of the male palp for fertilization.

The descendants of the Cyphophthalmi were, on the one hand, the Acari, on the other the Opiliones and the Ricinulei. The former developed their own peculiar type of penis, while the latter made use of their third legs to carry out the insertion of the sperm-packet. Lastly, the Solifugae used the chelicerae for the same essential purpose.

Contemporary with these changes in adaptive radiation, there have been, inevitably, many other specializations, such as the evolution of courtship, the spinning of webs and the secretion of pheromones. Nevertheless, in the three groups or sub-classes the course of evolution has taken roughly parallel courses, which might be diagrammatically represented thus:

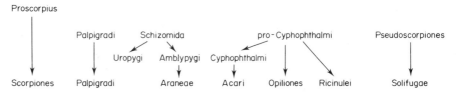

12

Taxonomy: Classification

It is only to be expected that the difficulties of determining the evolutionary history of the Arachnida will be reflected in the diversity of the schemes which have been proposed for their classification.

The earliest systems need not be considered here; the first that will be mentioned is that proposed by Ray Lankester in 1905, included in the tenth edition of the "Encyclopaedia Britannica", and widely adopted for some years. In this scheme the class was divided unequally into two grades, Anomomeristica and Nomomeristica, depending on constancy or inconstancy in the number of body somites. The former grade included only the fossil Trilobita. The Nomomeristica were also unequally divided into a sub-class Pantopoda, containing three orders of Pycnogonida, and a sub-class Euarachnida, containing the rest. In Euarachnida there were again two unequal divisions, the grade, Delobranchia or Hydropneustea for Limulus and the fossil Eurypterida, and Embolobranchia or Aeropneustea for the rest. Lastly, in Embolobranchia there were distinguished the section Pectinifera for the order Scorpiones and the section Epectinata for the remainder.

In 1904 Hansen and Sørensen suggested that the Pedipalpi, Araneida, Palpigradi and Ricinulei should be grouped together as Arachnida micrura, leaving the others in a second group, for which no name was proposed.

Lankester's method of attacking the class by successively cutting off the most aberrant group was preferred and in the circumstances was the best way in which sub-divisions could be produced—if such sub-divisions were to be regarded from the start as necessary or desirable.

The method used by J. H. Comstock (1944) is noteworthy only because it placed the Xiphosura in a separate but closely allied class, the Palaeostraca.

Lankester's system was retained with modifications in a classification offered by R. I. Pocock (1911). In this scheme Anomomeristica and Nomomeristica were preserved but the division of the latter into Pantopoda and Euarachnida disappeared. The Pycnogonida were excluded, and Nomomeristica were divided into four sub-classes, Limulava for the extinct Copura, Merostomata for Limulus and Eurypterida, Pectinifera for the scorpions, and Epectinata for eight other orders. In the Epectinata were six new divisions known as super-orders; one of these, Caulogastra, included Pedipalpi, Araneae and Palpigradi: the others contained but a single order each. The features of this classification were the expulsion of the Pycnogonida and the expression of the belief that the three orders of the Caulogastra are more closely related to each other than to any other Epectinata; and also that no two other orders of Epectinata are so closely related that they can share a super-order.

Both these systems placed the scorpions in a section or sub-class Pectinifera, which distinguished them from and contrasted them with all other living Arachnida. This opinion expresses the traditional view that the scorpions are more primitive arachnids than any others, and this, as has been seen, is scarcely true.

Neither Lankester nor Pocock had given much attention to the position of the extinct orders other than Eurypterida, and this subject was for the first time adequately considered by Størmer (1944). He named the phylum Arachnomorpha and divided it into two sub-phyla, Trilobitomorpha and Chelicerata. The latter contained two classes, Merostomata for Eurypterida and Xiphosura, and Arachnida for 13 orders, covering the 17 orders described in Part III of this book. No suggestion was made for the further relations of these orders, the names of which were simply recorded in a linear series.

This aspect of the classification was the outstanding feature of Petrunkevitch's study of palaeozoic Arachnida in 1949. As a result of detailed examination of a very large number of fossil Arachnida, he threw new light on the probable course of evolution and expressed his conclusions by placing the orders in four sub-classes which he named Latigastra, Caulogastra, Stethostomata and Soluta. At almost the same time the encyclopaedic "Traité de Zoologie" edited by Grassé, printed a straightforward list of 14 orders.

Petrunkevitch used his system, with minor changes in nomenclature, in his contribution to the "Treatise on Invertebrate Palaeontology", edited by R. C. Moore, and published in 1955. It was followed almost at once by a new classification, put forward by W. B. Dubinin in 1957.

This author was evidently in an iconoclastic mood and he suggested a drastic revision. The terrestrial Chelicerata were divided into four classes, containing 21 orders, clearly the outcome of an entirely fresh approach. More recently another arrangement was adopted by the present author in 1971. The three efforts may be compared in tabular form.

Petrunkevitch, 1955	Dubinin, 1957	Savory, 1971
Class Arachnida	Class Scorpionomorphes	Class Arachnida
Sub-class Latigastra	Order Eurypterida	Sub-class Scorpionomorpha
Order Scorpiones	Order Scorpionides	Super-order Scorpionides
Order Pseudoscorpiones	Order Pseudoscorpionides	Order Scorpiones
Order Opiliones	Order Palpigrades	Order Pseudoscorpiones
Order Architarbi	Order Uropyges	Super-order Rostrata
Order Acari	Order Kustarachnides	Order Solifugae
Sub-class Stethostomata	Order Amblypyges	Sub-class Stethostomata
Order Haptoda	Order Ricinuleides	Order Haptopoda
Order Anthracomarti		Order Anthracomarti
Sub-class Soluta	Class Arachnides	Sub-class Soluta
Order Trigonotarbi	Order Opiliones	Order Trigonotarbi
Sub-class Caulogastra	Order Phalangiotarbides	Sub-class Arachnomorpha
Order Palpigradi	Order Haptopodes	Infra-class Arachnoidea
Order Uropygi	Order Anthracomartides	Super-order Arachnoides
Order Schizomida	Order Trigonotarbides	Order Araneida
Order Kustarachnae	Order Liphistiomorphes	Order Amblypygi
Order Amblypygi	Order Arachnomorphes	Super-order Thelyphonides
Order Araneida	Order Mygalomorphes	Order Uropygi
Order Solifugae	Order Araneomorphes	Order Schizomida
Order Ricinulei		Super-order Kustarachnides
	Class Acaromorphes	Order Kustarachnae
	Order Acariformes	Infra-class Latisternoidea
	Order Parasitiformes	Order Palpigradi
		Sub-class Opilionoidea
	Class Solifugomorphes	Super-order Opilionides
	Order Solifuges	Order Architarbi
		Order Opiliones
		Order Acari
		Super-order Podogona
		Order Ricinulei

Here are three different attempts to find a satisfactory way of expressing on paper the relationships between 16 orders. There are points in all of them that may be criticized, and there are points of resemblance that are obvious.

Petrunkevitch's system was largely the result of prolonged study of the fossil species from palaeozoic strata. It is therefore based almost entirely on external features and depends chiefly on the breadth of the first opisthosomatic somite, which may or may not be reduced to a pedicel, on the difference between the number of detectable opisthosomatic somites and the primitive number 12, and on the configuration of the mouth parts.

The dependence on the pedicel brings together in the Latigastra orders that are as different in most other respects as are the Scorpiones and the Opiliones; and in the Caulogastra, the Solifugae and Ricinulei. This is a consequence, familiar to taxonomists in other classes, of basing a classification on an insufficient number of characteristics; for this may leave a residue of aberrant groups, to be conveniently consigned to a "dustbin" taxon. In this case the Caulogastra supply the accommodating receptacle.

Dubinin's system was intended to be a complete revision, which should take account of embryonic and ontogenetic development, as well as on morphology. This, it was hoped, would override the many apparent differences between the existing orders and replace a seemingly heterogeneous by a more homogeneous, not to say a more logical grouping. As a result, the four orders of spiders are, by implication, as different from one another as are the Ricinulei and Scorpiones; and the relationships between the Opiliones and the Acari, and between the Araneae and Amblypygi are hidden. Moreover, the concept of "class" and "order" has been wholly changed.

The third system tried to take account of as large a number of characters as was possible, but again was almost wholly concerned with externals. It represents no more than a step towards the system about to be suggested and used here. Its basic concept is that the orders may be brought together in groups or "convenient assemblages", each of which represents a distinct evolutionary pedigree, and it is based on the arguments following.

Any attempt to construct a classification of the Arachnida highlights the problem that faces every systematist. He has to choose whether he is going to devise a plan that will enable a taxonomist to place any given specimen in its correct group; or alternatively to produce an acceptable scheme that will show a phylogenist the course that evolution has followed as the different taxa have come into existence.

For breaking the class Arachnida into orders, the first of these choices is scarcely worthwhile. The orders are so distinct that they can be recognized at once. It would be very difficult to mistake a scorpion for a spider or a Ricinuleid for a mite, or a harvestman for a solifugid. The second choice presents a virtual impossibility, since the paths of arachnid evolution are virtually unknown, are obscure and are lacking in guiding principles. In such circumstances evidence must be drawn from details, the merest hint must be sympathetically considered, and, above all, speculation must not be forbidden.

There are two interlacing yet independent features of the orders of the Arachnida, which affect fundamentally any attempt to propose a logical order for their arrangement. These are, first, the sporadic distribution of certain characteristics among the orders. For example, (a) there are enlarged pedipalpi in scorpions and false scorpions, (b) a pedicel may form the first abdominal somite in Araneae, Amblypygi, Uropygi, Schizomida and Ricinulei, (c) the last three abdominal somites are compressed in Araneae, Palpigradi, Ricinulei, Amblypygi, Uropygi and Schizomida, (d) there is a terminal flagellum in Schizomida, Uropygi and Palpigradi, (e) there is a six-legged phase in Acari and Ricinulei.

Secondly, and in addition to this, the features emphasized as important in our diagnoses tend conspicuously to occur in alternative forms. A least ten may be mentioned.

(1) The carapace may be uniform or in part segmented.
(2) The opisthosoma may be uniform or segmented.
(3) The pedicel may be present or absent.
(4) The sternum may be uniform or segmented.
(5) The chelicerae may be of two or three podomeres.
(6) The pedipalpi may be leg-like or pincer-like.
(7) The coxae of legs or pedipalpi may bear gnathobases or may not.
(8) The first leg may be used as a leg or like an antenna.
(9) The legs may be of seven podomeres or may be sub-segmented anywhere.
(10) The coxae may meet and hide the sternum, or may be separated.

Here are ten pairs of alternatives, and ten alternatives may theoretically be arranged in 2^{10} or 1,024 combinations. Thus, even with a limit of ten pairs, we might expect 1,024 orders of Arachnida. But this is not so, and we may pause to ask why we have been robbed of 1,007 orders and left with a mere 17.

Three answers are possible. One is that perhaps many of the 1,024 have at some time actually come into existence, but were ill-adapted to survive and have left no trace of their short lives. Secondly, some of the combinations of chromosomes may have generated a lethal gene, and, like the homozygous yellow mouse, never have seen the light. Thirdly, some may be in existence, lurking unsuspected as sub-orders or super-families.

Be this as it may, there are 17 orders among which all the characteristics just mentioned are to be found. To the question "To what do these characteristics owe their existence?" the answer today will presumably be "To the presence of certain chromosomes, or to certain genes in the chromosomes, or to certain nucleotides in the DNA spiral". Since chromosomes separate, divide, cross-over and are variously distributed at meiosis, one is inclined, if not forced, to visualize groups, which may well be called stable gene complexes, each responsible for a particular characteristic in the adult arachnid.

There is much that this hypothesis can explain or make intelligible. For example, why do the Amblypygi seem to resemble the spiders while the Uropygi seem to recall the scorpions; why do pseudoscorpions merely look like scorpions while their true relations seem to be the Solifugae; why have the Uropygi a flagellum and the Amblypygi none and yet a short flagellum is found on salticid spiders of the genus Mantisatta; why is there a mysterious resemblance between the Cyphophthalmi, the Notostigmata and the Architarbi? Why do we never agree on an acceptable system of classification; and why are all the orders so different from one another?

Finally, where did all the orders of Arachnida come from? The answer is that their origin was enshrined in the gene pool drifting about the waters of the prehistoric oceans, carried in the gonads of the early Eurypterida. From here they have emerged in a set of combinations determined by chance.

Too much emphasis cannot be laid on the fortuitous nature of the whole process. This implies that the orders of Arachnida are the consequences of chance combinations of stable gene complexes, which were not only viable but were able to maintain themselves in the environment in which they were produced. Here is the basis of a hypothesis that may be described as aleatory evolution.

In the light of this hypothesis an attempt must now be made to arrange the orders of Arachnida in as logical a manner as is possible, to be tested by illustrating it in the form of a dendrogram (Fig. 27).

(1) A start may obviously be made with the Scorpiones, which are placed on the left of the diagram, as suitable to their primitive position. Their separation from the rest reflects their occurrence in the Silurian strata before any other order is known to have existed.

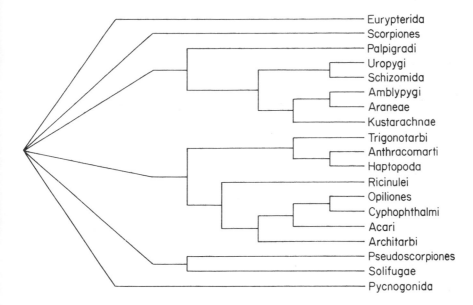

FIG. 27. Dendrogram of the orders of Arachnida.

(2) What I have called "the scorpion pattern" (Chap. 17) is repeated in the Uropygi. Based on an overall similarity, this is emphasized and confirmed by their general bionomics. The mode of life, the conduct of "bloodless battles" between half-hearted disputants, the character of the courtship and the type of spermatophore are all features which suggest that the "whip scorpion assemblage" should follow the Scorpiones. The first of these, however, must be the Palpigradi, which show a larger proportion of primitive characteristics than any other living order. Alone of this grouping they retain the primitive three-jointed chelcerae, but their terminal flagellum may be taken as evidence for their association with the Uropygi.

No relation is indicated between the Palpigradi and Scorpiones other than the features common to all Arachnida. The Schizomida and Uropygi are more closely allied, so that a common ancestry is shown, while the deeper origin of the Palpigradi denotes their phylogenic position.

(3) The Amblypygi manifestly follow. Though the flagellum, already shorter in the Schizomida, has disappeared, their relation to the Uropygi is beyond question. In other respects, the Amblypygi recall the

Araneae, and so support the idea that they show "the spider pattern". Their general appearance is that of a spider with long legs and no spinnerets; the shape of the sternum recalls that of the Liphistiidae, and the sub-pharyngeal part of the nervous system is striking evidence of this relationship.

The fossil Kustarachnae may claim the last place in this grouping. They show relationships to both the spiders and the Schizomida, but they appear to be older than either of these orders, a belief that is indicated by the deeper start of the line leading to their name.

(4) If the fact that an order which lived in the Carboniferous is now extinct may be taken to imply a failure to adapt to changed circumstances, then the Kustarachnae must be followed by other orders now extinct. Of these the Trigonotarbi, which alone appear in the Devonian, must take first place, with a deeper originating level. At that period they were well developed, and so, very possibly, were other orders whose remains have never been found.

They were not very distant relatives of the Anthracomarti, which therefore occupy the next place, and bring with them the other member of Petrunkevitch's Stethostomata, the Haptopoda. This order, in which the species had two eyes, 11 opisthosomatic somites, three-segmented chelicerae and sub-segmented tarsi; sufficiently recall the Opiliones to secure next place for this order.

(5) In this assemblage the Opiliones must undoubtedly be accompanied by the Acari, because the notostigmatic mites show such close relationships with the Cyphophthalmi; and with them must come the extinct Architarbi. This recalls Petrunkevitch's suggestion that "these three represent three branches of a single ancestral group".

In this possibly surprising trio there may be found the characteristics of blindness, a thickened exoskeleton with furrows marking the original segments, shortened abdomen without a telson, sub-segmentation of the tarsi, a cryptozoic life with indifference to starvation, colours unusual in the class, and above all, the existence of nymphs with six legs. This catalogue forms so good a diagnosis of the Ricinulei that it clearly points to the inclusion of that order in the next place.

(6) Two orders remain, the Pseudoscorpiones and the Solifugae. Many systematists have isolated the latter from all the others and put them in the last place in their classifications, but the realization that the false scorpions are only superficially like the real scorpions has led to the recognition of their relations with the Solifugae. To a superficial glance these two orders are not as similar as are the false and true scorpions, but other characteristics bring the Solifugae into the true perspective. Thus, both orders retain clear and complete segmentation of the abdomen, both lack pedicel, pygidium and telson; in both the pedal coxae

meet in the middle line, and both have two-jointed chelicerae. Their most obvious difference lies in the development of the chelicerae in the Solifugae and the pedipalpi in the Pseudoscorpiones, but these are adaptations to their different environments and modes of living. It may be added, for good measure, that both are reported to relish bed bugs in their diet.

The dendrogram most aptly raises the problem of the writing out of a classification of Arachnida in normal tabular form, for in doing this a decision must be made between two opposing principles.

The simple solution is that presented among the preliminary matter of this book (p. xi). The class is divided into four sub-classes, in each of which the orders are placed in a uniform list, following the succession of their names in the dendrogram. The result is a sort of index, conveying the minimum of information.

The alternative and less simple solution attempts to suggest the degrees of similarity and difference between the orders. This can be done by introducing a number of intermediate taxa between sub-class and order: in the classification given below these are called infra-classes, cohorts and super-orders. If these subsidiary taxa are to be given names some inventive thought is needed, and probably strict adherence to pure classical accidence is impossible. However, no arachnologist is obliged or expected to accept either the method or the nomenclature; for the devices of taxonomy have always been subject to individual opinions and the International Code of Zoological Nomenclature allows complete freedom of choice above the names of families. With these reservations in mind, the classification may be written as follows.

CLASS ARACHNIDA
    Sub-class Scorpionmorphae
            Order 1   Scorpiones
    Sub-class Arachnomorphae
      Infra-class Palpigradoidea
            Order 2   Palpigradi
      Infra-class Arachnoidea
        Cohort Uropygaceae
          Super-order Uropygoides
            Order 3   Uropygi
            Order 4   Schizomida
        Cohort Aranaceae
          Super-order Aranoides
            Order 5   Amblypygi
            Order 6   Araneae
          Super-order Kustarachnoides
            Order 7   Kustarachnae

Sub-class Opilionomorphae
  Infra-class Trigonotarboidea
    Super-order Trigonotarboides
      Order 8   Trigonotarbi
    Super-order Anthracomartoides
      Order 9   Anthracomarti
      Order 10 Haptopoda
  Infra-class Opilionoidea
    Cohort Ricinuliaceae
      Order 11 Ricinulei
    Cohort Opilionaceae
      Super-order Opilionoides
        Order 12 Opiliones
        Order 13 Cyphophthalmi
      Super-order Acaroides
        Order 14 Acari
    Cohort Arachitarbaceae
      Order 15 Architarbi
Sub-class Chelonethomorphae
      Order 16 Pseudoscorpiones
      Order 17 Solifugae

III. PROLES ARACHNES

13

The Order Scorpiones

[Scorpionides Latreille, 1817; Scorpiones Hemprich, 1826; Scorpi-
ides Koch, 1837; Scorpionida Pearse, 1936]

*Arachnida of considerable size in which the prosoma is covered by an
undivided carapace bearing two median and two to five pairs of lateral
eyes. The opisthosoma is markedly divisible into a mesosoma of seven (or
eight) somites with tergites that are broader than long, and a metasoma of
five quadrate or cylindrical somites, forming the "tail". There is no pedicel.
A telson is present in the form of a poison-bearing sting. The chelicerae are
small, of three segments, chelate. The pedipalpi are characteristically large
and powerful, and are of six segments, chelate. The legs are of seven
segments, without sub-articulations, and the tarsi carry two claws. The
sternum is triangular or pentagonal. The second opisthosomatic sternite
bears a pair of combs or pectines. Four pairs of book-lungs open on the
third, fourth, fifth and sixth sternites.*

The prosoma of the scorpions is uniformly covered with a hard cephalic
shield or carapace, its width equal to or even greater than its length.
There are no extensive traces of the primitive segmentation. There is a
median furrow in the middle of the carapace, running from behind the
eyes to the posterior margin where it widens to form a triangular
depression: this is an uncommon feature among the primitive orders of
Arachnida (Fig. 28).

The median eyes are placed close together on a low ocular tubercle
usually some way from the anterior border of the carapace, but genera
occur in which its position is either farther forward or farther back than
the normal. The lateral eyes form groups of two, three, four or five
small ocelli almost equal in size. A few scorpions are blind. Although
the eyes are outwardly similar they differ in development. The median
eyes are derived from two layers of hypodermis and are therefore

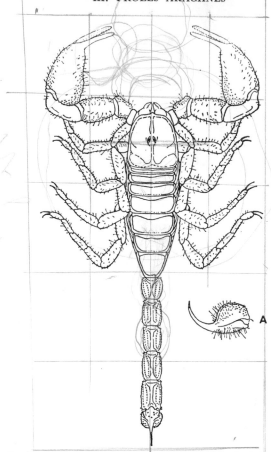

Fig. 28. A scorpion; dorsal aspect. Species, *Buthus occitanus*.

described as diplostichous; while the lateral eyes are monostichous and more closely resemble the eyes of Limulus.

The chelicerae (Fig. 29) are of three segments. The first of these is ring-like and is concealed by the edge of the carapace; the second is somewhat longer, convex above and outside, coated with setae on its inner surface and produced on this side into a pointed and toothed process. The third segment is the movable portion and is articulated to the second outside this process. Like the fixed process it is curved and toothed, but it is rather longer and ends in two points between which the tip of the fixed process rests. The teeth with which these parts are provided are much used in classification.

The pedipalpi are of six segments and are large efficient weapons, very characteristic of the order.

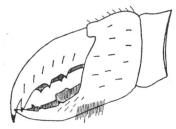

FIG. 29. Chelicera of Buthus, drawn from inside.

The coxa is almost cubical and possesses no maxillary process; the trochanter is also quite short. The third segment is long, and when at rest lies directed backward, parallel with the side of the carapace. The fourth segment lies at right angles to the third, pointing outward, and the fifth making another right angle, points forward. This penultimate segment is often very large; it is continued on its inner side into a pointed toothed process, against which the freely moving sixth segment closes, repeating the plan of the chelicerae on a much larger scale. The edges of the forceps are provided with pointed tubercles, which may be regular or irregular in size and disposition.

The chelicerae are, however, much more than mere weapons with which the scorpion attacks its prey or defends itself in combat. They are carriers of systematically arranged trichobothria, sense organs that have been exhaustively studied by Vachon (1973). His work, mentioned in Chapter 3, must be recognized as marking the birth of trichobothrio-taxy; and well illustrates the manner in which modern systematic arachnology progresses as more and more precise attention is given to matters of apparent detail.

The legs are of seven segments: the first pair are the shortest and the fourth the longest. The coxae are very large and in such close contact with one another that they form practically the whole of the lower surface of the prosoma. The two posterior pairs are immovable, but the first two pairs are movable and being provided with manducatory lobes form accessory mouth parts. The arrangement here is unique. The second coxa is the larger and more conspicuous; it is subtriangular in form, the base of the triangle lying to the outside, somewhat indented where the trochanter articulates with it. On the fore-edge of the coxa near the apex of the triangle is a strong forwardly-directed blade or apophysis which, at its distal end, meets its fellow in the middle line. The first coxa is the smallest. It lies in the angle between the second coxa and its process, and has a similar smaller process of its own which lies outside and close against that of the second (Fig. 30). Thus the

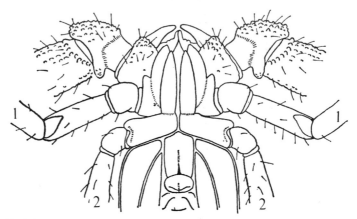

FIG. 30. Mouth parts of scorpion, showing the coxal processes on legs 1 and 2.

mouth is furnished with four blades from pedal coxae, which slowly triturate the food as it is held in the chelicerae.

The third coxae resemble the second in shape, but they have no apophysis and are about twice as large. The fourth are even longer: they do not broaden so much towards their distal ends, which reach past the centre of the first abdominal sternite.

The remaining segments of the legs are normal in form: their homologies were discussed above. The tarsi bear two large curved claws without teeth, and below and between them is a third median smaller claw. Some of the segments before the last may bear spurs at their distal ends.

The sternum of all scorpions is a very small plate between the third and the fourth coxae. In some genera it is only a narrow transverse strip of chitin, in others it is a small triangular plate and in the rest it is pentagonal, as in all young ones.

Close behind the sternum (Fig. 31) is the plate-like genital operculum, simple and inconspicuous, and immediately behind this, lying close to the fourth coxae, are the pectines. These are peculiar appendages, quite characteristic of the scorpions which have derived them from the first book-gills of the Xiphosura. Their use has never been quite satisfactorily described, and many different suggestions have been put forward. In the early literature of scorpion biology at least half-a-dozen ideas as to their function may be found:

(1) They detect the presence of food.

(2) They are used in mating to hold the male and female close together.

(3) They clean the body and limbs.

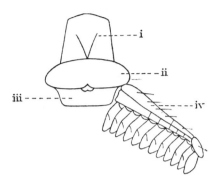

FIG. 31. Sternum and pectine of Euscorpius. (i) Sternum; (ii) genital operculum; (iii) sternite; (iv) pectine.

(4) They are accessory respiratory organs.

(5) They act as fans, driving air to the lungs.

(6) They are secondary sex organs.

Most of these are fanciful imaginations. More recent opinions are based on the fact that the pectines are plentifully supplied with nerves, indicating that they may be some kind of sense organ. Experiment seems to show that they have several functions:

(7) They react to the dryness or moistness of the air.

(8) They detect vibrations of the ground, giving warning of the approach of enemies or prey.

(9) They determine whether the ground is smooth enough or hard enough for the depositing of the spermatophore.

The back of the pectine is made of three pieces, the proximal part the longest and the middle one the shortest. The number of teeth is different in different species and varies from four to over 30.

The opisthosoma is manifestly divisible into a mesosoma and a metasoma, of seven and five somites respectively. The tergites of the first six mesosomatic somites are of gradually increasing length, the seventh is characteristically trapezoid in shape. The pleura between them become much stretched during pregnancy. Only five sternites are visible, the first four carrying each a pair of slit-like openings of the book-lungs.

The somites of the metasoma or "tail" are subcylindrical, the tergite and sternite of each being fused to form a ring of chitin. The upper side has a median groove, while the sides and lower surface bear a variable number of parallel longitudinal ridges of small spines. The last segment bears the telson, a bulb-like reservoir which contains the poison gland and which is produced into the sharp curved point of the sting (see A, Fig. 28).

DISTRIBUTION

Scorpions are found only in the warmer parts of the world (Fig. 32). In the northern hemisphere they occur in the countries bordering the Mediterranean, and are to be found in the south of France (five species) and in southern Germany. In America they reach the west coast, cross into Canada and are recorded from British Columbia, Alberta and Saskatchewan. In the southern hemisphere they are also widespread, but are absent from New Zealand, Patagonia and the oceanic islands.

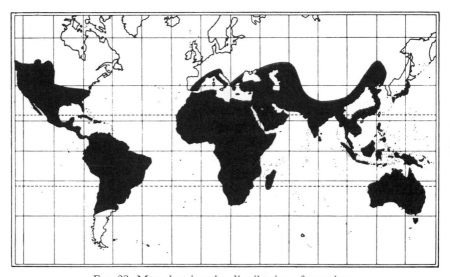

FIG. 32. Map showing the distribution of scorpions.

PALAEONTOLOGY

Fossil scorpions range from the Silurian to the Oligocene. The earliest known scorpion is the Isle of Gotland species, *Palaeophonus nuncius*, described by Thorell and Lindstrom (1885). This scorpion differed from all living forms in the shape of its tarsi, which were sharply pointed and possessed no terminal claws, a fact which has been taken by some to support the hypothesis that these Silurian scorpions were water-dwellers. In any case, Petrunkevitch has placed Palaeophonus in a separate sub-order, Protoscorpionina, which, besides the family Palaeoscorpionida also contains the families Dolichophonidae and Maxonidae, with three genera and three species from Scotland and North America.

Carboniferous scorpions, with claws on their tarsi and therefore placed in a sub-order named Euscorpionina, are fairly numerous, and occupy 22 genera with 38 species in six families:

Palaeoscorpiidae	Cyclophthalmidae
Archaeoctonidae	Isobuthidae
Eoscorpiidae	Centromachidae.

One of the most interesting of these is the species *Gigantoscorpio willsi* from Scotland, described by Størmer (1963). Its great size of 36 cm long is far in excess of that of any other known scorpion, and may be taken to suggest that an arachnid of such dimensions would need the support of water. In other respects the Carboniferous scorpions showed all the features characteristic of living species, and it appears that the order really reached its acme during this epoch. The specimens have been discovered in Bohemia, Britain, Pennsylvania, Illinois and New York.

The Mesozoic era has yielded remains of a family of Triassic scorpions, known as the Mesophonidae. This contains the genus Mesophonus, with six species, all from the Lower Keuper Sandstone of Lanark, and Spongiophonus, with one species from the Midlands. Berland points out that the rarity of Triassic scorpions is to be expected, for animals which prefer a dry habitat are not likely to die in circumstances favourable to preservation. This idea cannot, however, be made to account for the almost complete absence of fossil Arachnida from any other Mesozoic strata, and makes all the greater the interest of Mesophonus and the Palpigrade Sternarthron.

A few Tertiary scorpions have been found. One of these, *Tityus eogenus*, preserved in amber, is to be included in the recent family Buthidae, so marked is its resemblance to living forms.

Two considerations of more than usual interest have followed from the study of fossil scorpions.

In the sub-order Euscorpionina, mentioned above, the family Eoscorpiidae shared a super-family, Scorpionoidea, with the seven families of living scorpions. In other words, of all the Carboniferous families one only has survived in the form of scorpions of today. Eight other super-families have left no descendants, suggesting that during the period evolution and competition were rapid and intense.

Vachon, comparing the characteristics of the extinct families with the distribution of the existing ones, points out that the African deserts have not always been the arid, infertile tracts of the present time, but that they have undergone alternating wet and dry periods. Scorpions react to physical changes by adapting to specialized habits and restricted habitats, from which it again follows that existing scorpion fauna is to be regarded as a relict group, the survivors of inorganic evolution that destroyed the majority.

CLASSIFICATION

The classification of this order is based on the admirable foundations contained in "Das Tierreich'" of 1899 and due to Karl Kraepelin. He divided the living species into six families and his system has been slightly modified by later workers who have promoted his Diplocentrinae to the status of a family and have, generally, separated the Chaerilidae from the Chactidae.

The Buthidae are the largest family, with over 300 species and more than 30 genera. They are very widely distributed and are among the most venomous of their kind. Nearly half the American scorpions belong to this family. Most of these belong to the sub-family Centrurinae, which is typically western, but one familiar species, *Isometrus maculatus*, is now of worldwide distribution. As its name suggests it has a yellow body with a number of black spots, and is peculiar in that the tail of the male is twice as long as that of the female. It has travelled far, hidden on ships, and is even found on several islands where it has established itself. The other sub-family, the Buthinae, is mostly an Old World group, but Ananteris is a small American genus. The best known scorpion, *Buthus occitanus* (= *europaeus*), the common yellow scorpion of the Mediterranean, belongs to this sub-family, as does also *Microbuthus pusillus* from Aden, which is only 13 mm long and is the smallest scorpion.

The Scorpionidae are the second largest family, numbering some 150 species in about 20 genera. They are spread throughout the tropics and are toxic, though not so dangerous as the Buthidae. There are five sub-families. The Urodacinae include the Australian genus Urodacus, with only two lateral eyes, and the Hemiscorpioninae are another small group peculiar to Arabia. Most of the members of the family belong to the sub-family Scorpioninae, which is widespread and includes the largest living species *Pandinus imperator* and *P. gambiensis*. The Ischnurinae are found in both Africa and America and have sometimes been classified as a separate family, Ischnuridae.

The Diplocentridae are a small, mainly neotropical group, whose type, Diplocentrus, is Mexican and is also found in Texas and California. The genus Nebo contains large species from the western Mediterranean.

The Bothriuridae are mainly a South American family, but at least one genus, Cercophonius, is Australian. The typical genus, Bothriurus, is remarkable for a white patch of membrane on the poison reservoir, just behind the base of the sting. In this family the sternum is unusual in that it consists of two transverse plates, broader than long and sometimes inconspicuous.

The Vejovidae contain about 45 neotropical species, as well as the genus Jurus from the Mediterranean and Scorpiops from India.

The Chactidae are mainly an American family, but the well known black Euscorpius is found in France, Corsica, Italy and Algeria. It is sometimes transported accidentally to other parts, but unlike Isometrus fails to establish itself in distant regions. This family also contains *Belisarius zambeni*, the blind scorpion of the Pyrenees. The sub-family Megacorminae is typically Mexican. This family also includes Superstitiana, a genus from the south-western USA, remarkable for three stripes on its body, as well as Broteas, whose species, *B. alleni*, is barely 2·5 cm long and is one of the smallest scorpions.

The Chaerilidae are a small family of one or two genera, from tropical Asia.

The last six families are in many ways so distinct from the Buthidae that these have been considered as representing a distinct line of evolution within the order.

The families may be separated by means of the following key.

CLASSIFICATION OF THE ORDER SCORPIONES

1	(2)	Sternum longer than or as long as broad	3
2	(1)	Sternum of two transverse plates and therefore much broader than long, sometimes scarcely visible	BOTHRIURIDAE
3	(4)	Proximal end of each tarsus with one external spine; sternum clearly pentagonal	5
4	(3)	Proximal end of each tarsus with two stout spines; sternum often pointed anteriorly	7
5	(6)	No tooth or tubercle below the sting	SCORPIONIDAE
6	(5)	A tooth-like spine below the sting	DIPLOCENTRIDAE
7	(10)	Two lateral eyes on each side, rarely none	8
8	(9)	Pedipalp with three trichobothria on femur and not less than ten on fixed finger	CHACTIDAE
9	(8)	Pedipalp with nine trichobothria on femur and eight on fixed finger	CHAERILIDAE
10	(7)	Three to five lateral eyes on each side	11
11	(12)	Sternum triangular	BUTHIDAE
12	(11)	Sternum wide, with parallel sides	VEJOVIDAE

The scorpion has an evil reputation, which it does not deserve. One of the few invertebrates that have attracted attention from the earliest times, it has a place in the Zodiac and a gathering of legends and superstitious beliefs. Fear of the scorpion is traditional and is partly unfounded; but is not as foolish as the ideas that it does not drink, or that exposed to fire it commits suicide.

The biological interest of scorpions is enough to allow us to forget these absurdities, for to the zoologist the scorpion presents more than its fair share of problems. Its courtship, for long very imperfectly known, is peculiar, its embryonic development is remarkable, and its apparent rate of multiplication is so small that its survival is almost miraculous. The proverbial "chastisement with scorpions" is able, if nothing else, to reduce the zoologist to a state of fitting humility.

14

The Order Palpigradi

[Microthelyphonida Grassi, 1885; Palpigradi Thorell, 1888; Palpigradida Pearse, 1936]
(Micro whip-scorpions)

Arachnida of small size in which the prosoma is covered by a carapace of three pieces, and the opisthosoma is of 11 somites, the last three narrowed and bearing a long, jointed telson. There are no eyes. The chelicerae are of three segments and are chelate. The pedipalpi are of six segments and are leg-like, with terminal claws. The legs are variously sub-segmented and end in paired claws. There are no gnathobases on any of the coxae, and the mouth is characteristically placed on a short projecting rostrum. The sternum is of four or five sternites.

The prosoma of Palpigradi is protected by three chitinous shields. The foremost of these, the propeltidium, is the largest; it is an oval plate which covers four somites, extending from the chelicerae to the second pair of legs. It is followed by the small mesopeltidium, belonging to the fifth somite, and a somewhat larger metapeltidium, lying dorsally on the sixth somite. In this respect it will be seen that the Palpigradi support Hansen's conception of the primitive arachnid, possessing a "head" of four segments and a thorax of two, and show relations to Solifugae, in which three fused somites are followed by three free ones (Fig. 33).

There are no eyes on the carapace, but two seta-like structures exist near its anterior margin and are usually described as sensory patches.

The chelicerae superficially resemble those of Opiliones. They are composed of three segments. The first is a straight rounded piece, projecting forward from beneath the anterior end of the cephalic shield. The second is at right angles to the first and therefore points downward; it has a pointed and toothed process extending downward on its inner edge. Against this works the third segment, which is also pointed and toothed and joins the second outside its process. Thus the

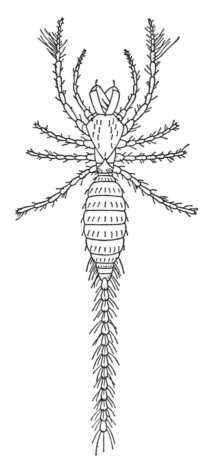

Fig. 33. Palpigradi; dorsal aspect. Species, *Koenenia mirabilis*.

chelicerae are chelate, the pincers working laterally. They are the only chelate limbs of Palpigradi and their only weapons; they are consequently capable of very free movement, and, with their long proximal segments, have a comparatively wide range (Fig. 34).

The pedipalpi are simple pedal structures, not specialized for any particular function and used in so leg-like a manner that, if the first pair of legs were used normally, Palpigradi might be said to possess five pairs of walking legs. The pedipalpi consist of nine podomeres, the tibia being followed by a basitarsus of two parts, while the tarsus is divided into three. There is a pair of terminal claws.

The first pair of true legs are by far the longest of the appendages, with a basitarsus (metatarsus) in four parts and a tarsus in three, it

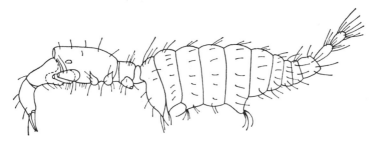

FIG. 34. *Koenenia mirabilis*; lateral aspect. After Kraepelin.

totals 12 pieces. The last six of these carry long sensory setae. The limb is not used for walking, but is carried stretched out ahead of the animal, in the same way as are the forelegs of some of the spiders and mites. It is clear that these legs are the chief sense organs of the animal. The second and third legs have the normal complement of seven segments each, but the fourth, owing to the possession of a divided tarsus, number eight segments. All the legs have two claws on the tarsi.

The mouth of Palpigradi is unique. Below the proximal end of the chelicerae is a soft egg-shaped prominence, pointing forward and composed of epistome and hypostome. The mouth appears as a transverse slit at the end of this oral process, lying as it were between the two lips. There is no other arachnid in which the mouth is so far forward as to lie between the basal joints of the chelicerae, and it may be that this is a primitive position, from which the mouth has not migrated as far back as it has in the allied orders.

The sternum is also very remarkable, differing from that of all other Arachnida and probably representing a much more primitive stage. There are four prosomatic sternites (Fig. 35). The anterior of these is the largest and consists of the fused sternites of the segments carrying the pedipalpi and the first pair of legs; the three succeeding ones are smaller and lie directly between their corresponding conical coxae. Thus if the hypostome be taken as representing sternite 1, the large anterior plate represents sternites 2 and 3, and the remainder sternites 4, 5 and 6. This is a full complement and there is no other arachnid prosoma with so well-defined a segmentation on its ventral surface.

A soft weakly-sclerotized pedicel unites the two parts of the animal's body, and is followed by an opisthosoma of 11 more somites. Tergites and sternites, though present, are weak and transparent; they are scarcely visible in ordinary circumstances, but can with some difficulty be seen with the help of a microscope. The genital orifice lies on

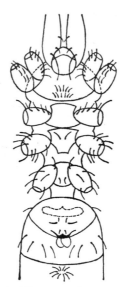

FIG. 35. *Koenenia mirabilis*; ventral aspect. After Kraepelin.

the ventral side of the first and second somites, and is of quite complicated appearance.

The second, third and fifth somites are characterized by single or
paired nipple-like humps on the ventral surface, set with strong sensory
setae. The last three somites are abruptly narrowed into a "tail",
similar to that of Uropygi and Ricinulei. The last carries a very characteristic post-abdomen, composed of 14 or 15 apparent segments, most of
which carry long setae. In life this curious organ is borne erect at right
angles to the body.

The first Palpigradi discovered were described as being without
respiratory organs. It was supposed that in such a small, soft animal
cutaneous interchange of gases would suffice. In the American species
belonging to the genus Prokoenenia paired lung-sacs have, however,
been reported in the fourth, fifth and sixth opisthosomatic sternites.
These sacs can apparently be evaginated by internal pressure and
invaginated by muscles, a pair of dorso-ventral muscles being attached
to each pair of lung-sacs.

DISTRIBUTION

First discovered at Catania in Sicily by B. Grassi in 1885, Palpigradi
have since been found at several places in Italy, notably among the
olive groves of Palmi in Calabria, in northern Africa and in France,

where they are not rare, at Banyuls-sur-mer. Probably they could be found at many places on the Mediterranean. Ten species have been described from Madagascar. In the East they have been recorded from Siam, in America from Chile, Texas and Paraguay, as well as from some of the northern states (Fig. 36). Although apparently feeble and defenceless, they seem to have powers of resistance which enable them to secure dispersal by the chance means provided by man. In 1914 Berland described a species which seemed to have become acclimatized in the Paris Museum, and in 1933 *Koenenia mirabilis* was found on the lower slopes of Mount Osmond, Adelaide, whither it had most probably been imported. These Arachnida are found under stones, often in company with Thysanura.

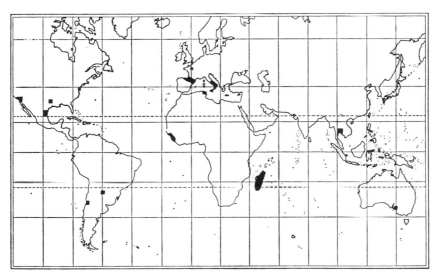

Fig. 36. Map showing distribution of Palpigradi.

CLASSIFICATION

Since 1885 the order has been held to consist of but a single family, the Koeneniidae. The discovery of other species, which now number about four dozen, has resulted in the formation of five genera.

The generic name Koenenia, used by Grassi in 1885, was in fact preoccupied, and in 1901 Borner split the genus into Eukoenenia and Prokoenenia. In 1913 Silvestri added two new genera for new species found in French Guinea; these were Allokoenenia and Koenenioides. In 1965 the interesting littoral genus Leptokoenenia was established by Condé.

The lowly phylogenetic position of the Palpigradi is strongly supported by the character of its solitary fossil representative, the Jurassic *Sternarthron zitteli*, described by Haase in 1890. Two specimens are preserved at Munich. This little creature had six separate prosomatic sternites (Fig. 37); if more specimens had been found and had been more fully described, Sternarthron would probably deserve a place in

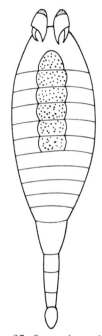

FIG. 37. *Sternarthron zitteli.*

palaeontology comparable to that occupied by such significant non-arachnid fossils as Palaeoisopus, Cladoselache or Cynognathus.

There is much in the structure of the Palpigradi that suggests a very primitive arachnid; yet there are, on the contrary, the rostrum, the sensory function of the first legs and the sub-segmentation of the hinder tarsi all of which are specializations. Even so the Palpigradi have a stronger claim to be considered as an order closer to the ancestral form of the Arachnida than Galeodes, Scorpio or any other group. All such claims are subject to the discount of some obvious specializations, such as inexplicable pectines, mysterious malleoli, silk-producing chelicerae or repugnatorial glands. But the body of a palpigrade surpasses them all in its possession of the following characteristics, all of which are examples of a primitive state or an early stage of evolution:

(i) segmented carapace	(iv) chelicerae of three segments
(ii) segmented abdomen	(v) non-chelate pedipalpi
(iii) absence of a pedicel	(vi) absence of gnathobases.

15

The Order Uropygi

[Thelyphoni Latreille, 1804; Thelyphonida Cambridge, 1872; Uropygi Thorell, 1882; Holopeltidia Borner, 1904; Thelyphonides Millot, 1942]
(Whip-scorpions, Vinegaroons; Vinaigriers; Geiselskorpione)

Arachnida in which the prosoma is longer than broad and is covered by an individual carapace, on which there are eight or 12 eyes. The opisthosoma is of 12 somites; the first very narrow and the last three reduced. A long whip-like jointed telson is present. The sternum is of three sternites. The first opisthosomatic sternite is very small, the second large, the third and fourth are both narrow. The chelicerae are of two segments, unchelate. The pedipalpi are of six segments, their coxae fused in the middle line and other segments with strong spine-like projections at their distal ends. The legs are of seven segments, the first pair with many sub-segments to the tarsi. Opisthosomatic glands secrete formic and acetic acids.

The prosoma of Uropygi is covered by a uniform carapace, almost rectangular in shape and longer than broad. Slight indentations, due to muscles within, can be seen on its surface. The eyes are in three groups. A pair of direct eyes occupies the centre of the carapace near its fore-edge; and on each side, at the level of the first coxae, there is a group of three or of five. Thus the total number is eight or 12.

The opisthosoma is a smooth elongated oval of 12 somites (Fig. 38). The first of these is reduced to form a pedicel and has only a very small and concealed sternite. Those following, from the second to the ninth, are all protected by transverse tergites. In some species these reach from side to side without a break, in others a median cleft divides the tergite into right and left halves. This division may be found in all tergites except the last three; or in the first three only; or again it may not divide the fourth, fifth and sixth sternites. The last three somites are much smaller; their tergites and sternites are united and they form

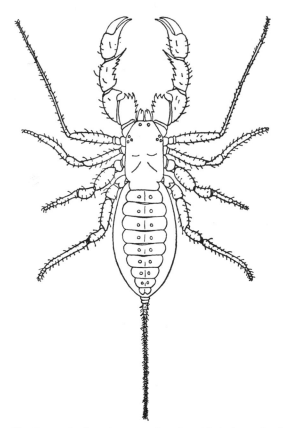

FIG. 38. Uropygi; dorsal aspect. Species, *Thelyphonus insularis*.

the post-abdomen or pygidium. On the twelfth opens the anus and the paired glands which, secreting formic and acetic acids, are the animal's very characteristic mode of self-defence. Equally characteristic, and responsible for the name of "whip-scorpions", is the long flagellum, which also arises from somite 12. It has many small segments, perhaps 30 or 40 in number, and is described as a telson. Its "segments" have therefore no homology with the somites of the body; as an anatomical feature its homologues are the sting of the scorpion and the very short flagellum of Schizomida.

The last or anal somite, from which the telson or flagellum arises, carries on its dorsal surface a pair of pale patches called ommatidia cr, better, ommatoids. They have not been shown to be sensitive to light and their function is unknown.

The comparative simplicity of the upper surface of the body is not to be seen below. The sternum is in three pieces; there is no labium, and the first visible sternite is a triangular piece lying between the first coxae and representing sternites 3 and 4. Behind it are two much smaller pieces, reduced in size by the approach of the coxae to the middle line.

On the lower side of the opisthosoma, the sternite of the second somite is very large. It carries the openings of the first pair of book-lungs, and the genital orifice is on its posterior margin. The second pair of book-lungs open on the third sternite; both this and the fourth are much narrower than the six which follow, and of which the first four are marked by muscular indentations.

The chelicerae are of two segments and are not chelate (Fig. 39). The proximal segments lie parallel to one another and have a certain

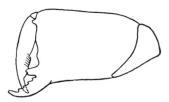

Fig. 39. Chelicera of Thelyphonus. After Kaestner.

degree of longitudinal freedom, as well as the ordinary ability to move sideways. They are thus far more loosely set than is customary. The second segment is a downwardly directed point of hard chitin, whose work is assisted by spines at the distal end of the first segment. These appendages therefore closely resemble the chelicerae of spiders, but they have no poison glands.

The pedipalpi are powerful appendages, composed of the usual six segments (Fig. 40). Their characteristic in Uropygi is the way in which their coxae meet and are fused together in the middle of the body, and thus they have no maxillary gnathites, very little freedom of movement and no masticatory function. But each trochanter has a large semi-circular process on its inner side, armed with several sharp tubercles. These processes can be pressed upon each other, or they can be opposed to the femora by bending the coxo-femoral articulation, and are thus well adapted to crushing the prey. The patella also has a conspicuous process, the "third apophysis", at its distal end on the inner side, against which the tibia closes, a second pincer. In some genera this patellar process is longer in the male than in the female. At the end of this appendage, the fixed finger, is a process of the tibia, and the movable finger is formed by fusion of the tarsus and basitarsus.

The legs of the first pair are not used for walking, but are held

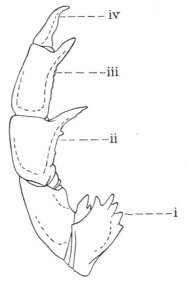

FIG. 40. Pedipalp of Thelyphonus. After Graveley. (i) Trochanter; (ii) tibia; (iii) transtarsus; (iv) tarsus.

stretched out in front as tactile organs. They are composed of seven true segments, but the tarsus is subdivided into nine parts. In the female these tarsal segments are swollen or variously modified. The true ambulatory limbs have the tarsus divided into three parts, ending in three claws.

The fusion of the pedipalpal coxae results in the formation of a buccal cavity in front of the true mouth or opening into the pharynx. In Uropygi this is formed by the palpal coxae below and at the sides and by the chelicerae and a small upper lip above.

DISTRIBUTION

Uropygi are purely tropical Arachnida, and are found in two separate areas in America, the southern states of the USA and the northeastern portion of South America. A few specimens of two species of the genus Hypoctonus have been taken in Gambia. They are not regarded as true African members of the order, but as immigrants that have been fortuitously introduced during the past century. They probably entered with the earth surrounding the roots of imported crop plants. Uropygi are also found in India and southern China, as well as in Malaysia and Japan (Fig. 41).

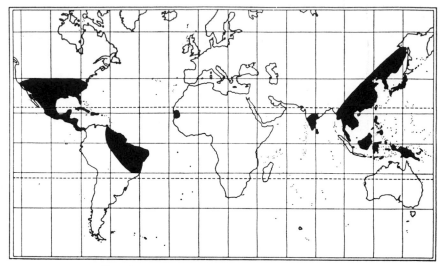

FIG. 41. Map showing distribution of Uropygi.

PALAEONTOLOGY

The fossil specimens of Uropygi emphasize the fact that this is one of the most unchanging orders of Arachnida, since they can all be placed in the same family, Thelyphonidae. There are two Carboniferous genera, Geralinura with two species from Mazon Creek, Illinois, and Prothely-phonus, represented by three species from Mazon Creek, Czecho-slovakia and Coseley, Worcestershire.

CLASSIFICATION

For many years all known species of this order were contained in one family, the Thelyphonidae, so named by Lucas in 1835. In a recent, valuable and very welcome revision by Rowland and Cooke (1973) the 85 species now known are placed in two families, the Hypoctonidae and the Thelyphonidae. Most of the former come from Malaysia, China and India. Among the latter, most of the typical genus Thelyphonus are found in the same region, while the sub-family Uroproctinae accommodates most of the American species, and the sub-family Typopeltinae is found in China and Thailand. The following key separates these groups.

The common name for Uropygi, "whip-scorpions", is very apt, for Uropygi illustrate an obvious modification of the scorpion-pattern. At the same time they have achieved so unusual an equilibrium with the

CLASSIFICATION OF THE ORDER UROPYGI

1 (2)	Keel between median and lateral eyes; ridge between median eyes	HYPOCTONIDAE
2 (1)	Keel between median and lateral eyes absent or very indistinct; no ridge between median eyes	THELYPHONIDAE (3)
3 (4)	Tarsus 1 of female normal; posterior abdominal tergites undivided	UROPROCTINAE
4 (3)	Tarsus 1 of female modified; posterior abdominal tergites divided	5
5 (6)	Patellar apophysis of male pedipalp normal; abdominal sternite 3 of female normal	THELYPHONINAE
6 (5)	Patellar apophysis of male pedipalp modified; abdominal sternite 3 of female greatly modified	TYPOPELTINAE

environment that they appear as the most stable order of Arachnida, and indeed as one of the most unchanging of all the orders of Arthropoda. There is an unusually close resemblance between the Carboniferous genera, Geralinura and Prothelyphonus, and the genera living today.

The Uropygi emphatically maintain what may well be called the arachnid tradition of long neglect in spite of their remarkable character. They are markedly cryptozoic, hiding in general in a moist habitat beneath the shelter of a stone or a decaying log. A few exceptional species, however, live in dry regions, and venture abroad at night during the rainy season. Their habits, therefore, have given them a false reputation for rarity, which is belied by the fact that in the right place and at the right season they can be collected in numbers. In these respects they recall the Ricinulei.

Had they no other peculiarity, their habit of secreting acetic acid would mark them off as a group that stands by itself, and it is a matter of great satisfaction that they have recently attracted from competent arachnologists the attention they so conspicuously deserve.

16

The Order Schizomida

[Tartarides Cambridge, 1872; Colopyga Cook, 1899; Schizopeltidia Börner, 1904; Schizomida Petrunkevitch, 1945]

Arachnida of small size, in which the prosoma is covered by a large propeltidium and smaller, paired, mesopeltidium and metapeltidium. There are no eyes. The opisthosoma is of 12 somites, of which the first is shortened and the last four are narrowed to form a pygidium, bearing a segmented telson. The sternum is represented by two triangular sternites between the coxae of the second and fourth legs respectively. The chelicerae are of two segments and are chelate. The pedipalpi are of six segments, unchelate, with pointed tarsi. Their coxae are fused to form a camarostome like that in the Uropygi. The legs are of seven segments, their tarsi with three claws. The first legs have no patella and are used as tactile organs.

The prosoma of Schizomida is covered by a segmented carapace of three portions, known as the propeltidium, mesopeltidium and metapeltidium. The foremost of these extends as far back as the second pair of legs and therefore corresponds to four original tergites. Its anterior margin carries between the bases of the chelicerae a small pointed tubercle, known as the epistome. The mesopeltidium consists of two small triangular transverse plates, and the metapeltidium, which is much larger, is also divided and has the form of two rectangular plates. There are no eyes, but on the propeltidium there are two smooth areas which are not lenses but which may represent the vestiges of vanished lateral eyes.

The chelicerae are of two segments and in position, form, and mode of action resemble those of spiders. They differ in having no poison glands and in having a strong projection on the inner side of the basal segment, so that the appendage is practically a chelate organ.

The pedipalpi are leg-like appendages of six segments. Though broader than the legs they are nothing like the powerful pedipalpi of

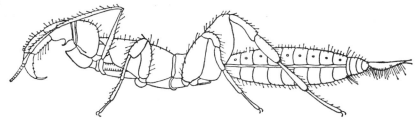

FIG. 42. Schizomida; lateral aspect. Species, *Nyctalops crassicaudatus*. After Pickard-Cambridge.

the Uropygi and form one of the most obvious differences between the two orders. The last segment is pointed, and is designated as an epitarsus rather than a claw. Unlike the corresponding organs of Uropygi these pedipalpi move vertically when in action.

The legs of the first pair are antenniform organs not used in walking. The coxae are unusually long; the tarsus is divided into eight pieces and there is neither epitarsus nor claw. The three posterior pairs of legs are more normal; they consist of seven segments and carry three terminal claws.

The opisthosoma is a smooth oval of 12 recognizable somites (Fig. 42). The first is reduced and forms the pedicel, retaining a small sternite. Somites 2 to 9 all possess tergites and sternites, united by pleural membrane. The sternite of the second somite is large, as in Uropygi, but that of the third is narrow. The last three somites are much constricted, forming the pygidium or post-abdomen, in which the unseparated tergites and sternites form complete hoops. The last somite bears the flagellum (Fig. 43), which in this order is short and consists of not more than four segments.

FIG. 43. Telson of Nyctalops. After Pickard-Cambridge.

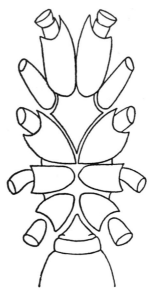

FIG. 44. Schizomida; ventral aspect. Species *Nyctalops crassicaudatis*. After Pickard-Cambridge.

The sternum in Schizomida, as in Uropygi, is represented by three separate sternites (Fig. 44).

DISTRIBUTION

The order is widely dispersed throughout the tropics. Its members are among those Arachnida that are able to survive accidental importations, and specimens have been recorded from Paris, Kew and Cambridge.

PALAEONTOLOGY

Fossil Schizomida are known only from the Tertiary. The first species was described by Petrunkevitch in 1945 and named *Calcitro fisheri*. Five specimens were found in Permian calcite from Ashford, Arizona. Species of two other genera, Calcoschizomus and Onychothelyphonus, were later described by Pierce from the same locality.

CLASSIFICATION

There is but a single family, Schizomidae or Tartaridae, with three genera, Schizomus, Trithyreus and Stenochrus. The last of these is known only from Puerto Rico.

When Millot in 1942 and Petrunkevitch in 1945 split the hetero-geneous order Pedipalpi, they both put the Schizomidae into an order of their own, but in Grassé's "Traité" in 1949 this order was re-united to the Uropygi as the family Schizomidae. They have been given separate treatment here, because the prevailing tendency in taxonomy today is towards separations, and it is probable that future opinion will confirm this step. The differences between Schizomida and Uropygi are at least as great as those between Uropygi and Amblypygi, including as they do both internal and external characteristics. The splitting of an order into two or three, or, what comes to the same thing, the promotion of sub-orders to the rank of orders, is not an uncommon occurrence in systematic zoology, and in the present instance it is well to note the distinctions on which the change has been based. A com-parison of this kind is best displayed in tabular form.

	AMBLYPYGI (Tarantulidae, Phrynidae)	UROPYGI (Oxypoei, Thelyphonidac)	SCHIZOMIDA (Tartarides, Schizopeltidia)
Carapace	Entire: broader than long	Entire: longer than broad	Divided into three regions
Eyes	Eight	Eight or 12	None
Opisthosoma	Twelve somites with no terminal flagellum	Twelve somites, 9–12 reduced, long flagellum	Twelve somites, 9–12 reduced, short flagellum
Chelicerae	Two segments	Two segments, basal with strong tubercle	Two segments, basal with strong tubercle
Pedipalpi	Strong, spinous raptorial	Strong, spinous, lateral movement	Leg-like, vertical movement
Legs 1	Very long and thin, tibia and tarsus sub-segmented	Long and thin, tarsus in nine pieces	Long, antenna-like, no claw
Legs 2–4	Normal, with pulvillus in some genera	Tarsus in three pieces, with three claws	Tarsus undivided, three claws

The species that belong to the three orders Pseudoscorpiones, Palpigradi and Schizomida are as different as any other three orders of Arachnida in body form and in the details of their appendages, but they have much in common in their general behaviour. Their superficial resemblances can generally be traced back to their small size. They are delicate animals, not strongly sclerotized; and all live their lives in well-hidden situations, usually in vegetable debris and nearly always near the ground. Here they all find a microclimate with a low evaporating power, an essential for nearly all of them. When taken out of their natural environments they show a general tendency to lose more water than they can afford and so to perish quickly in dry air. Here, too, they find a sufficient supply of the even smaller insects on which most of them prey.

There are species in the other orders which are no larger than are Schizomida and they are very often similarly constrained to live in regions of high humidity; but these three orders call our attention to the fact that one recognizable tendency among Arachnida is to evolve in the direction of small size and few degrees of freedom.

17

The Order Amblypygi

[Phryneides Gervais, 1844; Amblypygi Thorell, 1883; Phrynides
Millot, 1942; Phrynichida Petrunkevitch, 1945]
(Whip-spiders; Geiselspinnen)

*Arachnida in which the prosoma is broader than long and is covered by an
undivided carapace on which there are eight eyes. The opisthosoma is of
12 somites, the first forming a pedicel, the rest giving a smoothly rounded
outline to the body. The sternum is of three sternites. There is no telson.
The chelicerae are of two segments and are unchelate. The pedipalpi are of
six segments, large, powerful and spinous, unchelate, with the tarsi sharply
pointed. The legs of the first pair are very long, with sub-segmentation of
both the tarsi and the tibiae. They are not ambulatory. Legs 2 to 4 carry
pulvilli between the paired claws.*

The prosoma of Amblypygi is a smooth ovoid, broader than long and
covered by a uniform carapace. On this no trace of segmentation is to be
seen, but there is usually a median indentation surrounded by radiating
impressions, such as are a frequent characteristic of the carapace of
spiders. This resemblance, which runs through the whole of the
morphology of Amblypygi, is further emphasized by the eyes, of which
Amblypygi possess a direct pair in the middle near the fore-edge of the
carapace and two groups of three lateral eyes. The total, eight, is the
number most often found in spiders.

The opisthosoma is different in that it bears transverse chitinous
tergites which mark the original segmentation of the body. All 12
somites can be recognized; the first is reduced to a pedicel, the rest are
all distinctly marked by tergites and sternites united by pleural membrane; in fact the opisthosoma is one of the least altered by evolution
that is to be found among Arachnida (Fig. 45).

The lower surface of the prosoma shows the sternite of the original
third somite or tritosternum projecting between the pedipalpal coxae as

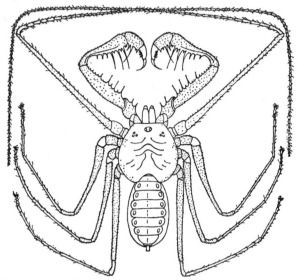

FIG. 45. Amblypygi; dorsal aspect. Species, *Stegophrynus dammermani.*

a pointed labium. Two more sternites lie behind it. The lower side of
the opisthosoma is characterized by the reduction of the last three
sternites: the small twelfth somite may be described as a pygidium.

The chelicerae are of two segments and are very spider-like both in
form and action (Fig. 46). There are, however, large pointed tubercles
on the first segment, which have no parallel among spiders, and there
are no venom glands within. In some species the basal segment carries
a group of short spines in the inner surface, forming with the opposing
group a stridulating organ.

The pedipalpi are the chief structural characteristic of the order.
They are of six strong segments with a pointed post-tarsus, and the
coxae, femora and tibiae are conspicuous for their length. The femora
and all following segments bear sharp pointed tubercles on their inner
sides, and the pedipalpi can thus be used to secure the prey. This can

FIG. 46. Chelicera of Stegophrynus. After Kaestner.

be done either by flexing one pedipalp against the victim, or by bringing both to bear on it from the sides. Even so, an insect thus seized sometimes manages to wriggle free; otherwise it is held by the pedipalpi while the chelicerae cut off pieces and masticate them below the mouth.

The pedipalpi have other functions. At the proximal end of the tarsus there are two curved rows of specialized setae of varying length, which form a cleaning brush, well adapted for use in the preening of the body and limbs. These setae are white or cream-coloured, not difficult to distinguish from the other "hairs", and those of the lower row are longer and more curved than those above them.

When two individuals meet and indulge in combat, they try to grasp each other with these appendages. The action is rapid, but no harm is inflicted, and the combatants soon separate.

The femur and tibia of the pedipalp are longer in the males than in the females, almost the only secondary sexual difference in this order.

The legs of the first pair are characteristically long and thin. They are not used in walking, but are stretched out in front of the animal as tactile organs. Tibia and tarsus are divided into a large number of subsegments, and the whiplash-like limb is generously supplied with setae. These first legs are exceptionally mobile. While the owner is at rest they are in constant touch with the surroundings, gently tapping the ground and anything thereon and bringing different impulses to the central nervous system. They recall the second legs of Opiliones, with their functions exaggerated and specialized.

All legs carry trichobothria on certain segments, and the long first legs are in particular the first to commence the vital operations of courtship. The legs of the second, third and fourth pairs are not as peculiar as the first pair. They consist of the ordinary seven segments followed by two curved claws, and in the family Charontidae a pulvillus as well (Fig. 47). Amblypygi can move very rapidly, and in the family just named can climb perpendicular surfaces and cling below horizontal ones.

Two pairs of book-lungs open on the second and third opisthosomatic sternites. They are of the usual type. In addition there are in some families two small ventral sacs near the middle of the third sternite. They are eversible under pressure from within, and perhaps have an accessory respiratory function.

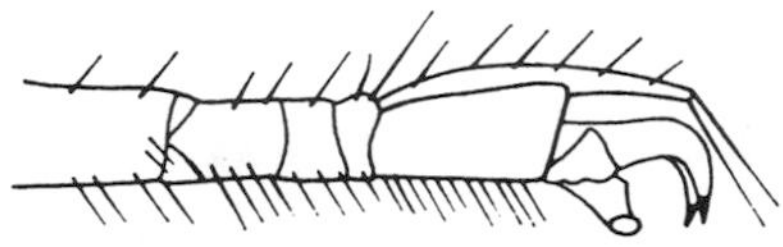

FIG. 47. Tarsus of *Charon grayi*. After Kraepelin.

DISTRIBUTION

Amblypygi occupy a continuous area in America, the greater part of which is south of Panama, are widespread in South Africa, and are found in India, Borneo and New Guinea. They do not, however, accompany Uropygi to the Asiatic coast (Fig. 48). Though a very large number of species have not been described, Amblypygi are often plentiful in the regions they inhabit: they live among rocks and in caves and are frequently found in houses.

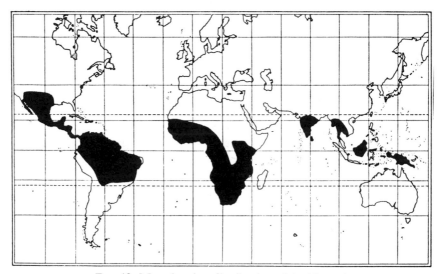

FIG. 48. Map showing distribution of Amblypygi.

PALAEONTOLOGY

The fossil record is meagre, for only three Carboniferous genera are known, Protophrynus, Thelyphrynus and Graeophyrynus, each of which is represented by an American species. In addition *Graeophrynus anglicus* was described by Pocock from Coseley, Worcestershire.

CLASSIFICATION

The order does not demand a very complex system of classification, for it is conveniently divided into two families, thus:

CLASSIFICATION OF THE ORDER AMBLYPYGI

1	(2)	Tarsi of legs 2 to 4 with pulvilli	CHARONTIDAE
2	(1)	Tarsi of legs 2 to 4 without pulvilli	TARANTULIDAE (3)

3	(4)	Tibia of leg 4 in one piece	PHRYNICINAE
4	(3)	Tibia of leg 4 in more than one piece	5
5	(6)	Tibia of leg 4 in two pieces	DAMONINAE
6	(5)	Tibia of leg 4 in three pieces	TARANTULINAE

If it may reasonably be said that Uropygi have maintained the scorpion-pattern, then it may be similarly suggested that Amblypygi have introduced the spider-pattern. In much of their structure the members of this order recall the order Araneae, and so closely that it is not over-fanciful to describe them as spiders without silk and perhaps as representatives of an ancestral stock from which spiders also evolved.

The mode of life adopted by Amblypygi is demonstrated by their generally flattened bodies. Nearly all Arachnida are inclined to rest in cracks and crevices, so that the flat bodies of Amblypygi should cause no surprise; a more conspicuous feature is their association with man. Other animals, better known because, perhaps, they are more unpleasant, have shown the same inclination; the cockroach and the rat are no doubt the most obvious. Amblypygi are less conspicuous, but are nevertheless sufficiently domesticated to cause Lawrence to say of *Damon variegatus* that "it probably occurs in all houses in Pietermaritzburg".

18

The Order Araneae

[Araneae Ovid; Aranei Clerck, 1757; Araneides Latreille, 1801; Araneidea Blackwall, 1861; Araneida Dallas, 1862]
(Spiders; Araignées; Spinnen)

Arachnida in which the prosoma is covered by an undivided carapace, usually with a transverse cephalic groove and carrying not more than eight simple eyes. The opisthosoma is unsegmented, its first somite being reduced to form the pedicel; pleopods of the fourth and fifth somites function as spinnerets. The anus is terminal and there is no telson. The chelicerae are of two segments, unchelate, and carrying a venom duct. The pedipalpi are of six segments, leg-like, tactile, and, in the males, carrying the male organs; their coxae carry gnathobases. The legs are of seven segments, their coxae without gnathobases, their tarsi with two or three claws. Respiration is by book-lungs or tracheae or both. The female orifice opens in a chitinous epigynum.

The prosoma of Araneae is protected by a uniform shield without much trace of segmentation beyond a groove, not always present, which separates a cephalic from a thoracic region. Upon the thoracic portions are usually indentations, a median fovea and eight radial striae, pointing towards the legs. These depressions mark the internal attachments of the muscles of the "sucking stomach" and the legs. The ocular region is sometimes darker than the rest, and the separate name of clypeus is usually given to that part of the prosoma between its anterior edge and the first row of eyes. The clypeus of spiders is never distinctly separated, but it corresponds to the cucullus of Ricinulei. Occasionally an elevation of the ocular region carries the eyes, or some of them, in a prominent position, sometimes exaggerated into a remarkable form.

The eyes are simple ocelli, the cornea of which is but a portion of the cuticle shaped to form a lens and free from setae and pigment. The majority of spiders have eight eyes but a number have six only. The genus Tetrablemma has four eyes and the genera Nops and Matta but two. Cave-dwelling spiders of the genus Anthrobia have no eyes. In many spiders it is obvious that the eyes are of different types, for some appear black and others pearly-white or yellow (Fig. 49).

The chelicerae consist invariably of two segments. The first, or paturon, is normally a stout conical part, which in a few families carries a smooth condyle articulated on the outer edge of its proximal end. This

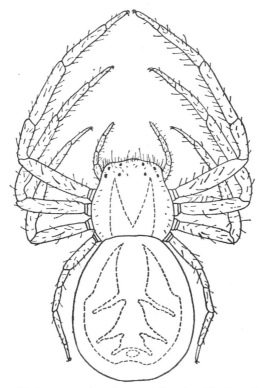

FIG. 49. Araneae; dorsal aspect. Species, *Xysticus cristatus*.

is something of a mystery; it might be a vestigial expodite. The outside of the lower edge is provided with a rake or rastellus of stout teeth in those families that dig burrows. The inner edge of the paturon sometimes bears a small nipple-like tuber, called the mastidion, of unknown function. The outer edge is sometimes corrugated by a series of ridges, which form part of a stridulating organ.

The second segment, or unguis, is a sharply pointed piece of very hard chitin, slightly curved. Its concave edge is grooved and finely toothed on the inside. Near the tip is the orifice of the duct of the venom gland. The chelicerae work in two different ways. Among the Liphistiomorphae and Theraphosomorphae they strike downwards and parallel to each other and are described as paraxial; among other spiders they strike transversely, tending to meet in the middle of the victim and are known as adaxial.

The pedipalpi consist of six segments (Fig. 50). Except in the more primitive sub-orders the coxa carries a maxillary lobe which has a compressing, masticatory function. The femur is sometimes used in stridulation. The remaining segments are used by young and female spiders as tactile organs, but in the mature male the tarsus is modified

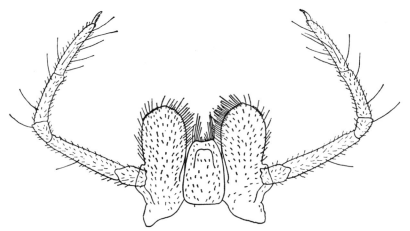

Fig. 50. Labium, maxillae and pedipalpi of a spider.

so that it can be used as an intromittent organ, carrying the sperm to the female spider. The male palpal tibia is often shorter than that of the female, and it often carries on its outer side a short process or apophysis which, during mating, is fitted into a groove in the female epigynum. This use of the pedipalp is one of several advances on the device of the spermatophore for insemination that occur among the Arachnida.

The male palpal organ lies near the tip of the tarsus in a cavity, the alveolus. In its simplest form it consists of a coiled tube, or receptaculum seminis, of three parts. These are a basal or proximal swollen bulb, the fundus, an intermediate reservoir, and, distally, a dark elongated ejaculatory duct.

The first advance from this simple condition is the migration of the whole genital bulb to the lower side of the tarsus and an increase in the size of the alveolus. The tarsus thus becomes more or less cup-like and is often renamed the cymbium. At the same time the palpal organ becomes divisible externally into three regions, containing the fundus, reservoir and duct respectively. The apical division is usually called the embolus.

In the next stage the embolus becomes divided into two. One of these is the ejaculatory duct or embolus proper; the other, called the conductor, protects it when the organ is at rest (Fig. 51). The tarsus, too, is sometimes divided into two parts, the smaller of which is called the paracymbium.

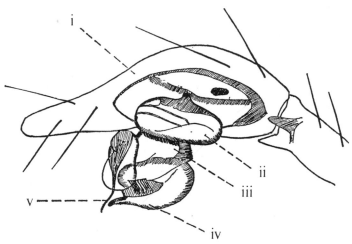

Fig. 51. Pedipalpal tarsus of male spider, *Tegenaria sp.*, showing palpal organ expanded. (i) cymbium; (ii) fundus; (iii) reservoir; (iv) embolus; (v) conductor.

In the most complicated type of palpus there are elaborations of these parts consisting chiefly in the addition of blood cavities or haematodochas, protecting rings of chitin and extra apophyses. Its numerous variations form a valuable means of characterizing the males of every species.

The legs are always composed of seven segments (Fig. 52). The coxae lie round the sternum and never exercise a masticatory function. The femora of some male mygalomorph spiders have small hooks on their inner surfaces; these are used in mating, when the male thrusts them against the chelicerae of the female, gagging her temporarily and reducing any possible risks to himself. The tarsi carry the paired claws which terminate the legs of all spiders. Sometimes an extension of the tarsus, the empodium or post-tarsus, extends between the paired claws, either as a pad, or as a bearer of adhesive hairs, or as a third, median, claw.

The claws, whether two or three in number, are hard and sharp, curved, and set with a row of teeth on the inside of the curve. Nielsen

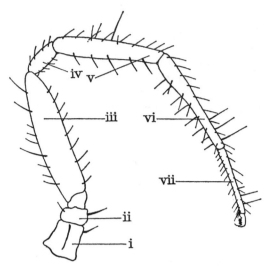

FIG. 52. Leg of spider. (i) Coxa; (ii) trochanter; (iii) femur; (iv) patella; (v) tibia; (vi) metatarsus; (vii) tarsus.

(1932) has described the action of the claws and claw-tufts on the threads of the web. The set of the claws gives a twist to the thread, and the claw-tuft forms a springy pad which releases the claws as the spider runs. In this unexpectedly elaborate way the spider avoids entanglement. The first leg is often but not always the longest, and only rarely is it held stretched out in front of the animal. In some male spiders it is decorated with tufts of setae or with black or coloured patches, which are displayed before the female during courtship.

The underside of the prosoma is formed from two unequal plates of chitin, the labium and the sternum. The former, which is the sternite of the second body somite, is situated between the maxillary lobes of the pedipalpi, and is variable in shape: square or elongated, semicircular or oval. Directly above it is a flattened cone of tissue, the rostrum. Below the rostrum is a chitinous plate, the epipharynx, and above the labium is a similar, corresponding plate, the hypopharynx. The epipharynx is marked with a longitudinal groove, the stomodaeum, up which the liquid food rises into the oesophagus, partly by surface tension and partly by the sucking action of the stomach within.

The sternum is generally oval or heart-shaped, but many variations are to be found in different families. It is slightly convex, and generally marked on each side by four bays or acetabula, which receive the coxae of the legs. There are also lyriform organs on its surface. The sternum probably represents the fused sternites of the third to the sixth somites, and Giltay has reported the existence in certain young specimens, suitably fixed, of three transverse striations, dividing it into the four regions expected. But this appearance cannot be seen in any adult.

The opisthosoma of spiders is normally a cylindrical or oval sac, with no outward sign of segmentation. Sometimes it has no pattern, but often there is a longitudinal mark above the heart, as well as small depressed points due to internal muscle attachments. Sometimes, however, there is an elaborate and even beautiful pattern. Segmentation persists in the sub-order Liphistiomorphae, where tergites indicate the existence of 12 opisthosomatic somites, while a smaller number of sternites protect the opisthosoma below. In some families an unsegmented dorsal plate is found.

The most remarkable feature of spiders' opisthosomas is the way in which they may, in some families, be strangely modified and developed into bizarre and fantastic forms. No other order of Arachnida shows anything like this, and the phenomenon cannot be readily explained. It may reasonably be said that a prickly abdomen is a discouragement to the tender mouths of predators, that a short rounded one may be helpful because the owner may be mistaken for a mollusc or a beetle, or that a long, thin one may mimic a caterpillar; but the peculiar shapes that are to be found seem to be merely awkward to control and valueless in themselves.

The underside of the opisthosoma shows more features than the upper. The region near the pedicel is more convex than the rest and is called the epigastrium: it is separated by the epigastric furrow. Two book-lungs, or the two anterior book-lungs of four-lunged spiders, lie in the epigastrium, and are conspicuous as pale patches. Behind them are either the posterior book-lungs or a pair of spiracles leading to tracheal tubes. In many species these spiracular tracheae lie at the ends of a transverse furrow, not easy to see and often far back, near the spinnerets.

The genital orifice lies in the middle of the epigastric furrow. In the male the vas deferens has a very small aperture, difficult to discern and unprotected by any epiandrium. The oviduct of the female has a larger aperture, in close association with the single or paired opening of the spermathecae, the whole surrounded by a complex epigyne (or epigynum) (Fig. 53). In this the actual vulva is protected by an operculum, the scape, and in the most elaborate forms there is a downward projection from the anterior side of the scape, called the crochet or clavus. This is occasionally accompanied by a second, posterior, process, the parmula: the two together act as a short ovipositor. Again, the unique character of Araneae is emphasized by the fact that in no other order is the female system provided with so elaborate an exit, which is different in every species, and, like its counterpart, the male palp, is the invariable method of identifying a species.

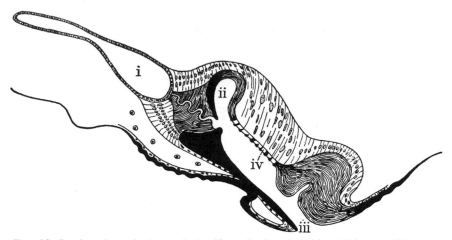

FIG. 53. Section through the genital orifice of a female spider, *Holocnemus hispanicus*. After Wiehle. (i) Uterus internus; (ii) receptaculum seminis; (iii) vagina; (iv) perforated plate of chitin.

In many spiders, but not in the Theraphosomorphae, nor in those possessing a cribellum, there is a small pointed appendage just in front of the spinnerets. This is the colulus. It is probably without function, being merely derived from the anterior median spinnerets by degeneration.

Behind the spinnerets a small tubercle, not always very obvious, carries the anus at its tip. This is sometimes called the anal tubercle, sometimes the post-abdomen. It is a vestigial structure, representing the remains of the last seven of the 12 opisthosomatic somites.

The only opisthosomatic appendages persisting in the adult spider are those of the fourth and fifth somites, where they function as spinning organs, namely, the cribellum, when this is present, and the six spinnerets (Fig. 54).

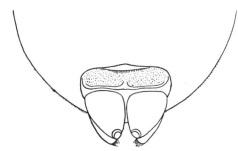

FIG. 54. Cribellum and spinnerets of spider, Ciniflo.

The number of spinnerets is not constant. In Liphistiomorphae the primitive number, eight, is found, occupying the middle of the ventral surface, but only the four expodites are said to be active. In most Theraphosomorphae there are two pairs only, belonging to the fifth somite, and in a few genera there is only one of these pairs. The relative lengths of the spinnerets in different families are also variable, and seem to be related to the method by which the silk is distributed. Where a sheet-web is made and the opisthosoma is swayed from side to side, the anterior spinnerets are very long.

The spinnerets are not the actual tubes through which the silk is secreted. The tip of a spinneret is covered with a number of minute tubes, through which the fluid silk passes. The smallest of these are called spools and are numerous. Those on the anterior spinnerets produce the attachment disc or transverse sweep of short threads which anchor a spider's line to the ground and those on the posterior spinnerets produce the broad ribbon or swathing band which is wrapped round a victim. The larger tubes or spigots produce the drag lines and foundation lines of the web; also the soft wadding found in egg cocoons and the viscid fluid which makes the spiral thread of a web adhesive.

The cribellum is an oval plate found just in front of the anterior spinnerets in certain families. It is perforated with a large number of minute pores, each of which is the orifice of the duct from a gland. Thus the cribellum produces a broad strip of silk composed of some hundreds of threads. This is combed out from the cribellum by the calamistrum (Fig. 55) on the fourth metatarsus and laid upon the plain silk strand which the spinnerets are simultaneously producing. The effect is to render the web more adhesive, or at any rate a harder entanglement from which to escape; it also gives it a characteristic bluish appearance.

The sense organs of spiders include the spines or setae with which their bodies and legs are covered, and the lyriform organs. Probably all the setae on a spider are more or less developed as sense organs, but some of those on the legs are useful accessories to the spinning organs. On examination it is easy to distinguish at least three different kinds. The most conspicuous are the stout sharp spines on the legs and pedipalpi, generally described as tactile. The most difficult to distinguish, even under the microscope, are the long delicate acoustic setae, believed to be receptors of sound waves. Intermediate between these extremes are many others, vaguely termed protective, and found in different forms, some club-shaped, some spatulate, some branched and some like smaller spines. In many cases spiders of a particular family are characterized by a special arrangement of leg spines (Fig. 55). The so-called acoustic setae are situated on the upper surface of the leg segments, either alone or in a series (Fig. 52).

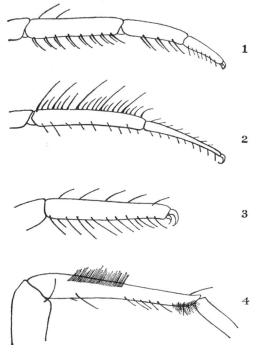

FIG. 55. Leg spines of spiders. 1, *Zora spinimana*; 2, *Ero furcata*; 3, tarsal comb of Theridion; 4, calamistrum of Ciniflo.

The lyriform organs are found on the legs, palpi, chelicerae and elsewhere on the body of spiders as well as on other Arachnida, but not on Acari, Solifugae or scorpions. They present the appearance of patches composed of ridges of chitin supplied internally with a branching nerve. These organs have been described in Chapter 3.

Many spiders are also provided with stridulating organs in different parts of the body. These consist in general of a tooth (or teeth) which rubs against a series of hard ridges and the relatively moving parts may be found on the prosoma and opisthosoma, on the chelicerae and pedipalpal femur, on the legs and pedipalpi or on the legs and pulmonary opercula. In many species these stridulating organs are confined to the male, being absent from or rudimentary in the female. Usually they do not produce a note audible to human ears, but there are exceptions to this and some spiders buzz like bees or purr like cats.

THE SILK GLANDS

The outstanding feature of the biology of spiders is the fact that their striking success may be traced to their exploitation of their silk-producing ability.

The uses which a spider makes of its silk are many, for unlike the silk of insect larvae it is used throughout life. Silk provides the gossamer threads on which young spiders migrate, webs are made of silk and captured insects are bound in silk fetters. Wanderers leave a silk drag-line behind them, sedentary spiders live in a silk home or in a silk-lined burrow, and eggs are laid in a silk cocoon. The silk products of spider industry may be classified as follows:

I. Linear structures
  1. The dragline.
  2. The gossamer thread.
  3. The webs of Miagrammopes and Cladomelea.
II. Plane structures
  4. The attachment discs which anchor threads.
  5. The swathing bands wrapped about insects.
  6. The similar sheets of the Theridiidae.
  7. The hackled band of cribellate spiders.
  8. The sperm web made by males.
III. Solid structures
  9. The webs of most spiders.
  10. The egg-cocoon.
  11. The nest or retreat.
  12. The moulting chamber.
  13. The mating chamber.
  14. The hibernating chamber.

The silk glands of spiders are situated in the posterior part of the opisthosoma and open by ducts at the spinnerets which are terminal except in the Liphistiidae. These glands were first investigated by Apstein (1889) and by Warburton (1890) working chiefly with Argiopidae, and were classified into seven types, according to their apparent functions and shapes. This classification, which has been widely quoted, was as follows.

  (i) Aciniform glands to median and posterior spinnerets.
  (ii) Pyriform glands to anterior spinnerets: these and the above produce the swathing bands and attachment discs.
  (iii) Ampullaceal glands to anterior spinnerets, produce all long threads.
  (iv) Cylindrical glands, peculiar to the female, produce the cocoon-silk.
  (v) Aggregate glands to posterior spinnerets, produce the spiral thread of orb-webs and its viscid coating.

(vi) Lobed glands, peculiar to the Theridiidae.
(vii) Flagelliform glands, which supply the core of the viscid spiral threads.
(viii) Cribellum glands, which feed the cribellum.

Attention may be directed to the flagelliform glands, not mentioned by Apstein or Warburton. They were discovered by Sekiguichi in 1952 and their existence was later denied. However, Anderson (1971) re-affirmed their existence, adding that they were "not very conspicuous". He also distinguished between large and small ampullaceal glands.

An investigation of these glands has been made by Millot (1929–31a). He found it impossible to retain this classification, for the glands are of many kinds and the types represented in one family are only rarely similar to those found in others. There is indeed a close similarity between all silk glands in histological structure and produce, but the form and disposition of the glands have been evolved independently in each group. There is no standard arrangement that can reasonably be described as the general type.

The nature of the silk glands and their relation to the poison glands is first suggested in the fact that the two kinds of gland occupy the same position in false scorpions and spiders. Their homology was proved by Millot's work (1929) on the spider Scytodes. Here the cephalothoracic gland is found not to be homogeneous throughout its length, but to consist of two kinds of cells. Those in the posterior lobe produce a silk-like glutinous substance, which the spider spits out at its insect prey, a method of attack which is unique among Arachnida. The cells in the anterior part are typical poison-producing cells, such as occur in the venom glands of all spiders. Individuals differ from one another in the proportion of the two kinds of cell present, with the result that the bites of some do not kill their victims, while the bites of others are quickly fatal.

This homology between silk and venom is maintained when we turn to the silk glands of Pseudoscorpiones. In this order the silk glands lie in the forepart of the prosoma, their ducts pass along the chelicerae and open at the so-called galea at the tip of the movable finger. Thus gland, duct and orifice of the silk-secreting apparatus of Pseudoscorpiones are completely homologous with the gland, duct and orifice of the poison-producing apparatus of Araneae.

Pseudoscorpiones use their silk solely for their own protection. They close the entrance to their hiding-places with a curtain and they build small nests of solid particles such as grains of sand, brushed over with layers of silk. In this nest they moult, incubate their eggs and hibernate.

The spinning activities of Acari have been summarized by André

(1932). The chief spinners are the Tetranychidae, which often live in colonies on trees and plants, and sometimes smother both sides of the leaves with layers of silk. Under this silk sheet they find shelter for themselves, their eggs and larvae. Many species are known from all parts of the world, and a single species does not habitually spin on the same kind of tree, but may be found on different species.

The silk of mites is so fine that a single thread is invisible to the naked eye. It is secreted by prosomatic glands, comparable to the silk glands of Pseudoscorpiones and the mixed silk and venom glands of the sicariid spiders. The ducts of the silk glands open inside the mouth, and by observation from beneath the liquid silk can be seen issuing therefrom. It is drawn out by the chelicerae, whose stylets or movable fingers manipulate it, sometimes assisted by the pedipalpi. The silk is never viscid.

In other families, a number of isolated examples of silk production have been reported, some of them with doubtful accuracy. Almost always the structure produced is a sheet-like shelter under which the mite moults or deposits its eggs. As examples there may be mentioned the species *Analges passerinus*, which spins nests on the bodies of birds under the feathers, where the nests may be found containing eggs and young; and *Oribata castanea*, which spins colonial tents under stones, sheltering 40 or more individuals.

DISTRIBUTION

The Liphistiomorphae are composed of two extinct families and one recent one. This includes but ten species, interesting because they are the most primitive living spiders (Fig. 56). Spiders of this sub-order were dominant in Palaeozoic times, but their survivors are limited to a relatively small area in eastern Asia (Fig. 57), where they live, some in caves, some on hills and some in the jungle. They make silk-lined burrows, closed with a trap-door.

The Theraphosomorphae include the true "trap-door spiders" or

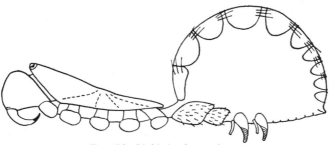

Fig. 56. *Liphistius batuensis.*

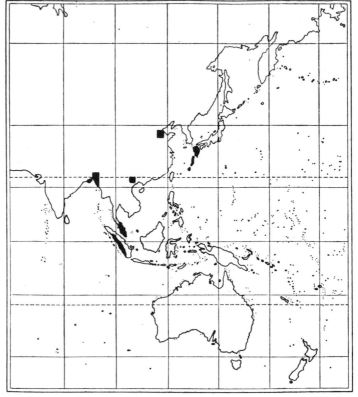

FIG. 57. Map showing distribution of Liphistiidae.

"tarantulas" of America. They are generally large spiders, with a squarish prosoma and eyes on a small ocular prominence. Their chelicerae, like those of the Liphistiidae, are articulated so that they strike vertically downwards, piercing their prey in parallel directions from above. The proximal segment is in many genera provided with a rake or rastellus used in excavating the burrow. The pedipalpi are long and leg-like and there are two pairs of book-lungs. In most genera there are four spinnerets. The sub-order includes wandering species, which hunt their prey, others which dig holes in the earth and close them with trap-doors, and others which spin silk tubes above the ground or make a web similar to that of Agelenidae.

The family Atypidae, which includes the British *Atypus affinis*, is the only one found in temperate regions. The Paratropididae from the Amazon and Pycnothelidae from Brazil are also small families. The rest are of wider range. The Migidae are found in South Africa, Madagascar and New Zealand. They do not burrow, but make perfectly concealed silk tubes on tree trunks. The Theraphosidae are the "bird-eating spiders", the typical hunters of the sub-order, and include the largest known species in the genera Theraphosa from New Guinea, Eurypelma and Avicularia from America.

The Labidognatha include the bulk of the world's spiders from all regions.

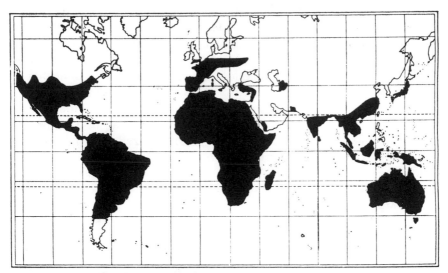

Fig. 58. Map showing distribution of Theraphosomorphae.

Among the cribellate forms, the Hypochilidae are most interesting. There are two species, *Hypochilus thorelli* from North Carolina and Tennessee, and *Ectatosticta davidi* from China. Their systematic position is doubtful, for they possess two pairs of book-lungs, a characteristic of the Theraphosomorphae, but in no other way are related to this sub-order. They have six spinnerets, and in many ways show primitive features. Some systematists have put them in a sub-order of their own.

The Amaurobiidae and Dictynidae are the most widespread representatives in temperate regions. The Eresidae have a quadrangular prosoma, very similar in shape to that of the Salticidae, and the Uloboridae are of great interest since they spin orb-webs.

The Apneumonatae include three somewhat aberrant families, Telemidae, Caponiidae and Symphytognathidae. The Telemidae are small, cavernicolous spiders. The genus Telema includes the blind species *T. tenella* from the Pyrenees; the other genus, Apneumonella, has eyes and lives in East Africa.

The Caponiidae include barely a dozen species. The genus Caponia is found in South Africa and Capomina in South America. From South America and some of the adjacent islands comes the genus Nops, peculiar in the possession of two eyes. The Symphytognathidae were founded on the remarkable spider *Symphytognatha globosa*, described by Hickman from Tasmania in 1931. In this spider the pedipalp of the female consists solely of the coxal segment with its gnathobase, a reduced condition almost without parallel in Arachnida.

Among the Dipneumonatae the families of the Sicarioidea include two which have six eyes, the Dysderidae and Oonopidae. These are relatively primitive families, whose web, if they make a web, is a bell-mouthed tubular retreat. The Dysderidae have a peculiar sternum, which overlaps the coxae of the legs, and the Oonopidae often have a hard plate or scutum, covering the dorsal surface of the opisthosoma.

In the large cohort Argiopoidea the Argyronetidae include but one species, the water spider, *Argyroneta aquatica*, familiar in Europe and northern Asia. The Hahniidae are characterized by the peculiar arrangement of their spinnerets which form a transverse line across the hind-edge of the opisthosoma. The Anyphaenidae are a large family, so are the Pholcidae; the others are all small and of local or very discontinuous distribution.

The Dionycha are in general spiders which wander in search of their food.

Five families are large and of cosmopolitan distribution. Of these the Drassidae and Clubionidae represent the wandering mainly nocturnal species, the Thomisidae and Sparassidae are flattened "crab-spiders" which lurk hidden in crevices or, concealed by protective colouration, lie in flowers waiting for visiting insects, and the Salticidae are large-eyed jumping-spiders, particularly numerous in the Tropics, which can see and leap upon their victims at a distance of several inches. The other families are only small groups.

The Trionycha include the peculiar Archaeidae, in which the caput is much raised in front and the chelicerae articulate far above the pedipalpi. Living forms are rare but exactly similar curiosities have been found "fossilized" in Baltic amber.

The Mimetidae are well characterized by the spinal armature of their legs, the Hersiliidae by their long tail-like spinnerets and the Pholcidae by their elongated narrow bodies and very long legs.

Three families of this group are huntsmen, the Oxyopidae, Pisauridae and Lycosidae, which pursue their prey and trust to speed to catch it. The last two are numerous and very widely distributed.

The Agelenidae are a large family, spinning the ordinary cob-web, a tubular resting-place the lower edge of which is continued into a wide hammock, held by threads above and below.

Finally there are the three largest families: the Theridiidae, characterized by a comb of setae on the tarsi of their fourth legs, small spiders spinning an irregular maze of threads among leaves; the Linyphiidae, including an enormous number of tiny spiders, often without pattern, which spin a sheet of web and live upside-down, hanging from its lower surface; and the Araneidae (—Epeiridae—Argiopidae) whose web is the familiar orb-web. All these spiders are spread throughout the world.

PALAEONTOLOGY

The Palaeozoic Araneae, of which at least 18 species have been described, are of interest because no fewer than 12 of them belong to the sub-order Liphistiomorphae. This sub-order, now limited to ten species in Malaysia, China and Japan, was apparently the dominant type of spider in those remote times, and was widely spread over the northern hemisphere. These 12 species have been placed in two wholly extinct families, the Arthrolycosidae, with two genera and three species, and the Arthromygalidae, with nine genera and nine species. Protolycosa, the first of these to be described, came in 1860 from Kattowitz in Upper Silesia, two Arthrolycosa from Illinois, and Eoctenizia from Coseley, Worcestershire. Most of the Arthromygalidae were described from Nyran, Bohemia, in 1888 or 1904.

Three members of the Arachnomorphae have been found in Palaeozoic strata, Eopholcus and Pyritaranea from the Carboniferous of Bohemia and Archeometa from Coseley. No Palaeozoic Mygalomorphae have been found.

Records of Arachnida from Mesozoic rocks are rare and only four or five species have so far been discovered. All came from the oolitic limestone of Papenheim, Bavaria.

Tertiary formations, including amber, have yielded much more. Theraphosomorphae are represented by Eoatypus from the Eocene of the Isle of Wight, and Eodiplurina from the Oligocene of North America. The sub-order Arachnomorphae is represented by fossils from the Miocene of Rott, Germany, and Aringen, Switzerland, and from the Eocene of Aix, the Isle of Wight and Florisant, Colorado. The Carboniferous type of Liphistiomorph is found not to have persisted in Europe.

The richest source of Tertiary spiders is Baltic amber, in which many invertebrate remains have been found. These are treated in a separate section in Part V.

Nearly 100 fossil species of Arachnomorphae have come from sources other than Baltic amber: most of these have been found in the Oligocene of Colorado, but France, Germany and Switzerland have also provided examples.

There are still no records of fossil spiders from the southern hemisphere.

CLASSIFICATION

The classification of spiders has always presented the difficulties that are provided by any large group of animals in which the number of known species and genera is rapidly increasing, and in which the course of evolution is so obscure that it can be traced only in outline.

Many systematists have agreed in recognizing three sub-orders, the first or most primitive of which is the Liphistiomorphae, containing the Asiatic Liphistiidae. The second sub-order contains the Theraphosomorphae, previously known as the Mygalomorphae: the character of these two groups is outlined above. All other spiders were placed in the third sub-order, Arachnomorphae or Araneomorphae or Gnaphosomorphae.

The breaking up of the second sub-order into eight or nine families has not proved to be difficult; not so the many families in the third sub-order. It has not proved easy to construct a truly phylogenetic system, so that schemes proposed have been purely artificial ones, based on external characteristics, and valuable chiefly because it made easier the identification of any specimen under examination. This is the practical value of any system of classifying objects of any sort.

One classification stands out above all the others, that of Petrunkevitch in 1933, a courageous attempt to include internal structures among the characteristics used.

As a result of cutting sections of all representative species, Petrunkevitch based his first divisions on the number of ostia through which the heart communicated with the pericardium. There is no doubt that the basic idea that all the organs of an animal should be considered in making a natural system of classification is theoretically so sound as to be beyond criticism; it is equally undoubted that very few practising systematists have both the time and the skill needed to dissect or to section all the animals they are asked to name, and as a result Petrunkevitch's system has been universally admired and simultaneously neglected.

Other systems have in general been attempts to improve on their predecessors: all made use of external features easily visible—the presence or absence of a cribellum, the simplicity or complexity of the sex organs, the existence of two or of three claws on the tarsi—and it be readily understood that the systems suggested have become ever more complex.

In the latest system, that of Bonnet (1959), in the second volume of his

"Bibliographia Araneorum", there is a summary of the characteristics of its predecessors, criticizes them where criticism is necessary, and adds that, with the objections now clearly seen, "c'est le fruit mûr qui ne demande qu'à être cueilli".

The result contains nine taxa above the genus, namely sub-order, legion, sub-legion, super-cohort, cohort, sub-cohort, super-family, family and sub-family. It also gives a number of new names to some of the taxa which do not come under the notice of the International Rules of Zoological Nomenclature, and which were revised for various reasons, some of them merely linguistic.

Bonnet's is the most elaborate and detailed classification of spiders as yet published, and would probably have received universal acceptance and approval had it not appeared at the end of his encyclopaedic Bibliographia Araneorum, whose circle of readers could not be a wide one. In outline it was as follows:

Sub-order Mesothelae
    Legion Liphistiomorphae   2 families
Sub-order Opisthothelae
    Super-legion Orthognatha
      Legion Theraphosomorphae   9 families
    Super-legion Labidognatha
      Legion Gnaphosomorphae
        Sub-legion Cribellatae   12 families
        Sub-legion Ecribellatae
          Super-cohort Apneumonatae
            Cohort Telemoidea   3 families
          Super-cohort Dipneumonatae
            Cohort Sicarioidea   6 families
            Cohort Argiopoidea
               Sub-cohort Trionycha   16 families
               Sub-cohort Dionycha   12 families

This scheme embraces a total of 60 families, to which several newly established families may be added. These include the Toxopidae, Gradungulidae, Textricellidae, Micropholcommatidae and Austrochilidae from the southern hemisphere; some new families produced by a splitting of the Sicariidae, and at least seven fossil families from amber.

KEY GIVING PARTIAL SEPARATION OF THE ABOVE GROUPS

1	(2)	Spinnerets in mid-ventral region of opisthosoma (Mesothelae)	LIPHISTIOMORPHAE
2	(1)	Spinnerets at posterior end of opisthosoma (Opisthothelae)	3
3	(4)	Chelicerae striking vertically and parallel (Orthognatha)	THERAPHOSOMORPHAE

4	(3)	Chelicerae striking transversely	
		(Labidognatha)	GNAPHOSOMORPHAE (5)
5	(6)	Cribellum present	
		(Cribellatae)	7
6	(5)	Cribellum absent	
		(Ecribellatae)	11
7	(8)	Four book-lungs	
		(Paleocribellatae)	HYPOCHILOIDEA
8	(7)	Two book-lungs	
		(Neocribellatae)	ULOBOROIDEA (9)
9	(10)	Three tarsal claws	PERISSONYCHA
10	(9)	Two tarsal claws	ARTIONYCHA
11	(12)	No book-lungs	
		(Apneumonatae)	TELEMOIDEA
12	(11)	Two book-lungs	
		(Dipneumonatae)	13
13	(14)	External genitalia simple	
		(Haplogynae)	SICARIOIDEA
14	(13)	External genitalia complex	
		(Entelegynae)	ARGIOPOIDEA (15)
15	(16)	Three tarsal claws	TRIONYCHA
16	(15)	Two tarsal claws	DIONYCHA

There must be a number of arachnologists who think that Bonnet's system is more detailed than is necessary for general purposes or for elementary study. In such circumstances a preferable, simplified, scheme is the division of the order into five sub-orders; Liphistiomorphae, Therphosomorphae, Hypochilomorphae, Apneumonomorphae and Gnaphosomorphae.

It would be both optimistic and unwise to suggest that the problems of spider classification are near solution. In the meantime, great interest attaches to the efforts of Lehtinen to construct a comparative and phylogenetic system. The publication of his first instalment (1967) which dealt with the cribellate families and their close relatives made manifest the immensity of his task, and forewarned his readers of the need for patience before the final tabulation was set before them.

The spider is the dominant arachnid; it surpasses all others in the number and variety of its species, in the complexity of its habits, and in the breadth of its range across the world. So well is it advertised by the beauty of the orb-web that all men know the spider, and for many it represents the whole class of Arachnida. The spider may be encountered in mythology, in history, in art, in literature; its reputation is not unspotted and its merits are seldom recognized. In particular, the organization of this book may seem to devote to the order Araneae fewer pages than it would seem to deserve, but the proportion is justified because the large number of available books on spiders far exceeds that on any of the other orders. At the same time this imbalance indicates the many opportunities that await the arachnologists who are attracted to a specialized study of the other orders, or of any one of them. A comprehensive understanding of the whole class can be expected only when knowledge of every order, large or small, blatant or obscure, is reasonably adequate.

19

The Order Kustarachnae

[Kustarachnae Petrunkevitch, 1913; Kustarachnida Petrunkevitch, 1955]

*Carboniferous Arachnida in which the prosoma is broader than long and is
covered by an undivided carapace, with two eyes on an ocular tubercle. The
opisthosoma is segmented and has a short terminal pygidium. The chelicerae
are unknown. The pedipalpi are of four segments and are chelate; their
coxae are completely fused in the middle line, concealing or obliterating
the labium. The sternum is very small and is centrally placed between the
triangular coxae. The trochanters of the legs are divided into two sub-
segments; the other segments are long and slender.*

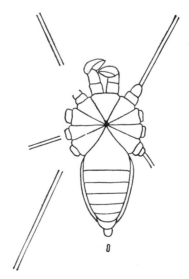

Fig. 59. *Kustarachne tenuipes.* After Petrunkevitch.

This is one of the smallest orders of the Arachnida, containing only one family, the Kustarachnidae, with one genus, Kustarachne (Fig. 59), and three species, each represented by one specimen from Mazon Creek, Illinois.

The peculiar fusion of the pedipalpal coxae implies the existence of a camarostome, and to this extent relates the order to the Schizomida.

20

The Order Trigonotarbi

[Trigonotarbi Petrunkevitch, 1949; Trigonotarbida Petrunkevitch, 1955]

Devonian and Carboniferous Arachnida, in which the prosoma is covered by an undivided carapace, bearing two eyes or none. Sternum and labium are present. The opisthosoma is joined to the prosoma either by the middle portion only or by its whole breadth, and is composed of eight to 11 somites, with an anal operculum. The tergites are divided into three plates by two longitudinal lines; the sternites are not so divided. The chelicerae are of two segments and are unchelate. The pedipalpi are of six segments, with no gnathobases on their coxae. The legs are of seven segments, with two tarsal claws.

The body of Trigonotarbi is unusual, and is subject to more variation than is the rule among the extinct orders of Arachnida, especially in the number and arrangement of the opisthosomatic somites (Fig. 60). The last tergite, whatever its number, may not be subdivided like the others into three plates, but, surrounded by the tergite before it, passes round the end of the body to a position on the ventral surface; here it lies next to the last sternite, between the sides of the sternite in front.

The shape of the carapace is also variable. It may be smoothly rounded in front, or it may be remarkably produced into a forwardly directed point, on so large a scale that in some genera the carapace is triangular. A further peculiarity is the existence of two types of junction, either complete or median only, between prosoma and opisthosoma, alternatives which are unknown in other orders.

The order has been split into five families:

Palaeocharinidae with 11 opisthosomatic somites
Anthracosironidae ,, 10 ,,

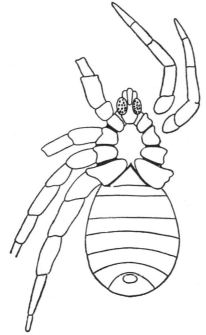

FIG. 60. *Trigonomartus pustulans*. After Petrunkevitch.

Eophrynidae   with   9 opisthosomatic somites
Trigonotarbidae   ,,   8   ,,
Trigonomartidae   ,,   8   ,,

The Trigonotarbi are an order of unusual interest, and probably merit the creation of a sub-class Soluta for their sole occupation (Petrunkevitch, 1949).

Occurring in the Devonian, they appear to be older than the other extinct orders, with the consequence that they have had opportunity to evolve a set of peculiarities which distinguish them clearly from these. Most obvious are the types of junction between prosoma and opisthosoma and the number of opisthosomatic somites.

The junction, as mentioned above, may extend across the whole body-breadth or it may be limited to the central fraction, which is, however, distinct from the narrow pedical of the Caulogastra. The number of opisthosomatic somites varies from 11 to eight. In some genera the last two are narrowed to form a pygidium, reminiscent of the three-somite pygidium in some living orders. In some genera the last tergite is turned under the body, where it is surrounded by the terminal sternites, a condition comparable to that in many Opiliones. Again, the number of book-lungs varies from two to four or even eight, as in the family Eophrynidae. It is this repeated instability in such basic structural formations that suggested the name Soluta for the sub-class.

Evidence of relationship between arachnid orders is often sought by examining the coxo-sternal arrangement, which plays so important a part in feeding. In the Trigonotarbi there is a labium anterior to the sternum, which is edged on each side by five coxae, a typical pattern shown also by the Araneae, but lacking the gnathobases of the pedipalpi. And a detail that also recalls some spiders of the family Araneidae is the existence of sharply pointed tubercles on the hind edge of the abdomen.

Trigonotarbi are also related to the Anthracomarti (Chapter 21), but can be distinguished from that order by the possession of three, not five, parts to the opisthosomatic tergites.

The general character of the order may be summarized by the statement that to a number of peculiarities of their own they add an almost equal number of resemblances to several other orders, a condition that is to be found elsewhere in the Arachnida and may be associated with a long period of evolution through the geological ages.

21

The Order Anthracomarti

[Anthracomarti Karsch, 1882; Anthracomartida Petrunkevitch, 1955]

Carboniferous and Permian Arachnida, in which the prosoma is covered by an undivided carapace, on which there are no eyes. The sternum is long and narrow, and there is no pedicel. The opisthosoma is of ten somites, completely represented by ten tergites and ten sternites; the first tergite may be connected to the carapace, the second and third are fused together. Tergites 2 to 9 are divided into five plates by four longitudinal lines, and the corresponding sternites are divided into three plates by two lines. The chelicerae are of three segments, the third being a retrovert fang. The pedipalpi are of six segments. The legs are of seven segments, with two claws and an onychium.

This order consists of a single family only, the Anthracomartidae, of 11 genera and fewer than 20 species, described from North America, Belgium, Britain, Czechoslovakia and Germany (Fig. 61). It is clear that in their day they were a fairly specialized type, for the mouth has moved a long way back from its primitive apical position. This may have necessitated a peculiar diet or a peculiar method of feeding.

The division of the tergites into five portions is most unusual, and it is difficult to imagine how such a condition came into existence, or what advantages it brought, save that a swelling of the abdomen after a large meal would have been made easier. The family seems, however, to have reached an advanced condition and one of comparative stability, so that no more than minor changes, resulting in the appearance of new genera, were possible.

An unusual feature was the possession of three pairs of book-lungs, opening on abdominal somites 2, 3 and 4.

The two small orders Haptopoda and Anthracomarti were placed together by Petrunkevitch in 1949 in a sub-class which he named Stethostomata. Their common feature, which distinguishes them from

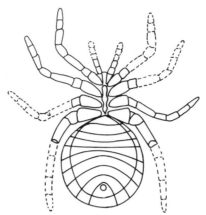

FIG. 61. Cryptomartus. After Petrunkevitch.

other Arachnida, is the unusual way in which the first or basal segments of the chelicerae are placed between the coxae of the pedipalpi, so that all four appendages originate at the same level. At the same time, and as a result, the mouth is moved to a position immediately in front of the sternum, where it therefore lies between the coxae of the first legs. There is little doubt that the various configurations of the mouth parts as they occur in the orders of Arachnida, as, indeed, elsewhere among the Arthropoda, often provide evidence of evolutionary relations.

The sub-class Stethostomata is a particularly interesting one for this reason and bears all the appearance of an unsuccessful combination of genes. One cannot but regret the fact that today we have no chance of seeing one of the Stethostomata consuming its food. A special kind of food may have been necessary, imposing an unwelcome limitation on the animal's everyday life. If environmental conditions changed these difficulties may have increased, and may have been among the causes of extinction.

22

The Order Haptopoda

[Haptopoda Pocock, 1911; Haptopodida Petrunkevitch, 1955]

Carboniferous Arachnida in which the prosoma is covered by an undivided carapace, bearing two eyes close to the fore-edge. The sternum is long and narrow and is of three sternites. The opisthosoma is of 11 somites, the first tergite being very short and wholly or partly hidden by the carapace. The first sternite is triangular, between the coxae of the fourth legs; the second is very large, convex behind, with two genital orifices. Sternites 3 to 10 are longitudinally divided by a median line. The chelicerae are of three segments, chelate. The pedipalpi are of six segments and have no gnathobases. The legs are of seven segments, with their tarsi sub-segmented, and are of relative lengths 1.4.3.2.

This order, founded by Pocock in 1911, contains but one family, the Plesiosironidae, with one genus, Plesiosiro, and a single species, *P.*

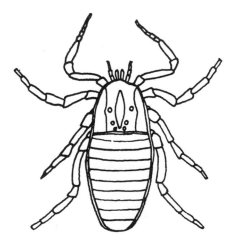

FIG. 62. *Plesiosiro madeleyi.* After Petrunkevitch.

madeleyi (Fig. 62). The animal was about 11 mm long. Nine specimens are known, all from Britain, where it was discovered at Coseley, Worcestershire. Therefore it was probably a common species, and one of the curiosities of arachnology is the fact that no specimen has been found on the Continent, and no other species of the order has occurred in any other part of the world.

23

The Order Architarbi

[Phalangiotarbi Haase, 1890; Architarbi Petrunkevitch, 1945; Architarbida Petrunkevitch, 1955]

Carboniferous Arachnida in which the prosoma is covered by an undivided carapace, having two eyes level with its surface or with an eye-tubercle, carrying three pairs of eyes. The opisthosoma is of ten or 11 somites; the first four to six somites are compressed and divided by a median line. There are usually nine sternites, similarly divided into three by two longitudinal lines; the first sternite is triangular and lies between the fourth coxae, the second has two orifices, the last is circular. The sternum is long and narrow and consists of three or four sternites. The chelicerae are of three segments and are chelate. The pedipalpi are slender and are of six segments, their coxae are situated above the coxae of the first legs, their gnathobases forming a stomotheca. The legs are of seven segments.

The chief peculiarity of this order is the compression of the anterior somites of the opisthosoma, so that their tergites are reduced to narrow strips. This shortens the body, a process which is also seen in Opiliones, where it has gone further. In fact there are several ways in which the Cyphophthalmi of Opiliones and some of the Acari are similar, but this is not certain evidence that they have been derived from a common ancestor. In a class in which, like Arachnida, so many alternative characters exist there are bound to be a number of purely fortuitous resemblances with no phylogenetic significance at all.

Architarbi are also remarkable in having a peculiar three-lobed ocular tubercle which carries three pairs of eyes, as well as in the division of the sternites into three regions by two longitudinal lines or furrows. This apparent fragmentation of either tergites or sternites is found in other extinct orders such as Trigonotarbi, but seems not to have persisted into recent times.

The order consists of three families.

The family Heterotarbidae contains one genus and one species, *Heterotarbus ovatus*. This species, about 14 mm long, has very small chelate chelicerae. Its pedipalpi are invisible, save for their trochanters which separate the first pair of legs. The first legs are long and slender, the rest short and stout. The opisthosoma consists of ten somites, five short followed by five long ones. The coxae and chelicerae of this species place it between the Opiliones and the rest of its own order.

The family Opiliotarbidae, founded by Petrunkevitch in 1949, contains only the species formerly known as *Architarbus elongatus*, described from Braidwood, Illinois, by Scudder in 1890.

The family Architarbidae contains 11 genera and some 20 species. The genus Geratarbus contains two American species, *G lacoei* and *G. minutus*, which have a small oval sternum and the first pair of coxae touching each other. The genus Discotarbus consists of one American species, *D. deplanatus*, which has a triangular prosoma and a wide rounded opisthosoma. Its sternum is divided into three areas. The genus Metatarbus, which also consists of one American species, *M. triangularis*, resembles the latter, but has an elongated oval abdomen. The only European genus is Phalangiotarbus, with one species, *P. subovalis*. It differs from its American allies in having a straight posterior edge to the prosoma instead of a procurved margin. In this family the coxae of the first pair of legs are contiguous throughout their entire length, and hide the chelicerae completely (Fig. 63). The genus Architarbus has a remarkable prosoma, pointed at both ends. The anterior margin is drawn out to a sharp point, almost like a spine. The posterior projection has

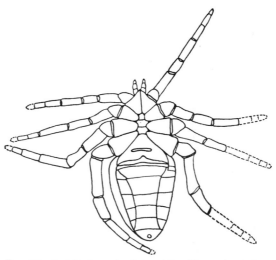

Fig. 63. *Architarbus rotundatus*. After Petrunkevitch.

curved sides meeting at a rounded point which somewhat displaces the anterior somites of the opisthosoma. This part of the body is rounded and the first somites are very short. This is a relatively large genus, with both American and European species.

This is the fifth and last order of extinct Arachnida and at this point it is of some interest to compare them thus:

	FAMILIES	GENERA	SPECIES
Kustarachnae	1	1	3
Trigonotarbi	5	13	20
Anthracomarti	1	10	16
Haptopoda	1	1	1
Architarbi	3	14	19

Thus the total number of described species is less than 60, and almost one-third of these belong to Architarbi. Their distribution included both European and American sites, but little can be deduced from this since the discovery of fossil invertebrates is more influenced by opportunity than any other factor.

All families were in existence during the Carboniferous, Trigonotarbi are also known from the Devonian and only Anthracomarti seem to have survived into the Permian.

24

The Order Opiliones

[Phalangidea Leach, 1815; Holetra Latreille, 1817; Opiliones Sundevall, 1833; Phalangida Perty, 1833; Phalangiida Petrunkevitch, 1955]
(Harvestmen; Faucheurs; Kanker, Afterspinnen, Weberknechte)

Arachnida in which the prosoma is covered by an unsegmented carapace, with two transverse grooves representing the two posterior somites. There are two eyes, rarely sessile, usually on the sides of an ocularium. The orifices of a pair of odoriferous glands open above the second coxae. There is no pedicel. The opisthosoma is of not more than nine somites, with an anal operculum. The sternum is absent or hidden. The chelicerae are of three segments and are chelate. The pedipalpi are of six segments with or without a tarsal claw; their coxae bear gnathobases. The legs are of seven segments, usually long, sometimes very long, their tarsi often with sub-segmentation and very flexible, ending in a claw or claws; the first two pairs of coxae carry gnathobases. The male has a protrusible penis.

The prosoma of Opiliones is in general uniformly covered by the fused tergites, but in the sub-order Laniatores the two posterior segments can be distinguished owing to transverse grooves. The carapace is usually smooth and is generally continuous with a dorsal shield over the opisthosoma.

Near the front edge are the eyes. In the majority of Opiliones the eyes are situated on an ocularium. This is a prominent tubercle, placed farther forward in Laniatores than in most of the Palpatores and carrying an eye on each side. The top of the tubercle is often spiny and its shape and size are very different in different families. In Trogulidae, the anterior border of the prosoma is produced forward into two curved processes covered with spines and carrying the eyes (Fig. 64). The under-surfaces of these projections are concave and form a hood over the chelicerae and pedipalpi.

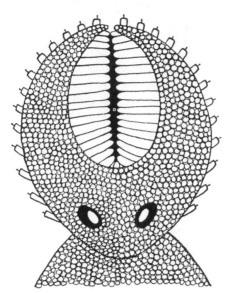

FIG. 64. Hood of Trogulus. After Kaestner.

The structure of the eyes in Palpatores has recently been investigated by Curtis (1970). He had described the lens as a clear, transparent region of cuticle, and below it is a layer of columnar lentigen cells, which has a "glassy" appearance. A thin pre-retinal membrane lies between this layer and the retina, which is composed of a number of retinulae, each composed of a cylindrical rhabdome tapering to an axon which forms part of the optic nerve. A layer of pigment cells outside the retina shuts off the lateral rays of light. The eyes of seven Phalangiidae studied have essentially the same construction, and those of Nemastoma are degenerate with shorter lentigen cells and irregular rhabdomes. This may be correlated with their living in a less well-illuminated habitat.

Curtis has also found an unexpected effect of prolonged darkness. After 24 hours in complete dark there is a disruption of the microvilli which surround the rhabdomes; they are replaced by spirally arranged lamellae and vesicles. This adaptation to low illumination is known to occur in some other invertebrates, but apparently not in so pronounced a manner as in the phalangiid eye.

The prosoma also bears a feature characteristic of the order, the orifices of a pair of odoriferous glands. In Palpatores these orifices are above the coxae of the first pair of legs and are clearly visible in many genera, although less conspicuous in others. In Laniatores the orifices are above the second coxae and are usually narrow oblique openings. In Cyphophthalmi they are situated at the ends of small conical processes, the coni foetidi, between the second and third pairs of coxae, just behind the eyes. In the past the apertures of these glands have been mistaken for spiracles and the glands themselves, which are often visible through the carapace, for eyes. Hence Cyphophthalmi were originally described as having stalked eyes. The glands secrete a fluid when the animal is irritated. In many species the odour of the fluid is not intense, and it has been variously described.

Simon (1879) wrote "chez les Phalangium opilio cette odeur rapelle celle de brou de noix"; Rossler (1882) mentioned "an aromatic odour"; Bristowe said of Gonyleptidae that they produced a "strong rather sweet smell"; and others have compared it to horseradish. Stipperger (1928) described "an offensive odour, lasting for about two minutes", while Blum and Edgar (1971) found that the secretion of two Leiobunum species contained 4-methyl-3-heptanone.

The chelicerae are of the three segments, the first projecting forward and the second downward at right angles to it. The second segment is produced on its inner edge and the third is articulated outside this process, forming a chela. These parts are pointed and finely toothed. In Palpatores the chelicerae are not as a rule large, though they may be somewhat elaborated in males (Fig. 65), but in many genera they are conspicuously carried, stretched out in front and having sharp spines on their inner surfaces.

The pedipalpi are of six segments (Fig. 66). The coxae bear gnathobases which form part of the rather complex mouth: the other segments

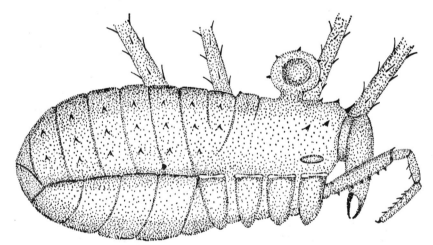

Fig. 65. Phalangium, lateral aspect.

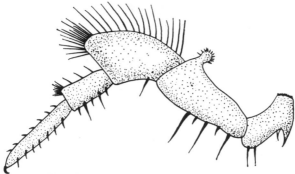

FIG. 66. Pedipalp of Megabunus.

are different in the three sub-orders. In Palpatores the appendage is relatively slender and almost always spineless. The tarsus usually carries a claw, which when present is almost immovable and is quite short, scarcely exceeding in length the width of the palp. The tarsus is longer than the tibia when it bears a claw and shorter when it does not. In Laniatores the pedipalpi are strong and armed with spines. The claws are large and can be closed against the tarsus so that they act almost like chelate organs.

The legs are of seven segments (Fig. 67). The first two pairs of coxae generally carry gnathobases, but in some families these are borne by the first pair only. The remaining segments are in some families no longer than the legs of spiders, but in the majority they are characterized by their extreme length, which indeed forms by far the most obvious feature of most of the familiar Opiliones. In Laniatores and Palpatores the legs of the first pair are the shortest, and the longest are the second pair in Palpatores and the fourth in most Laniatores. In this sub-order these posterior legs are often remarkably strong and stout with sharp spines on the femora. The tarsal segments are many-jointed, the number of joints ranging from a few to a hundred or more. In the latter case the tarsus becomes a delicate whip-like segment. In all Palpatores the tarsi carry one claw; in Laniatores the two anterior pairs of legs, have also one claw, while the two posterior pairs have two claws, sometimes pectinate. As a rule the tarsal claws in this order are simple.

The mouth of Opiliones lies between an epistome above and a labium below. The epistome projects from below the front edge of the carapace, and sometimes, as in Nemastomatidae and others, is preceded by an extra piece, the pre-epistome. The sides of the mouth are guarded by the maxillary gnathobases of the pedipalpi and first legs, with the second coxae sometimes as accessories behind them.

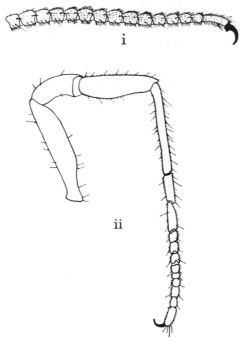

FIG. 67. Legs of Opiliones. After Kaestner. (i) Tarsus of *Opilio parietinus*; (ii) whole leg of *Nemastoma bimaculatum*.

The lower surface of the prosoma is formed from the labium, sternum and the coxae of the legs (Figs 68 and 69). The labium, as mentioned above, may be regarded as one of the mouth parts. It is probably homologous with the corresponding labium of Araneae and represents the sternite of one of the original somites. Behind it is the sternum proper, a single plate. In Laniatores it is long and narrow, reaching as far back as the fourth coxae. In Palpatores it is always short; sometimes it is continuous with the labial portion and sometimes with the genital plates behind it, but it is often largely out of sight, hidden below the coxae. The coxae are sometimes all immovable and sometimes all movable in Palpatores; in Laniatores the three posterior pairs are immovable, the fourth pair coalescing with the opisthosoma, while the first pair can be rotated, moving their maxillary lobes towards or away from the mandibles.

The opisthosoma is usually so closely united to the prosoma that the distinction between the two parts is not obvious. Its precise structure is not easily elucidated without the comparison of many forms, since segments are often fused together or are missing, and also because the

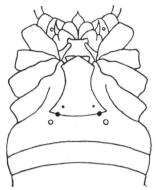

FIG. 68. Phalangium, ventral aspect. After Roewer.

tergites and sternites of the same segment are not always placed opposite each other in the lateral aspect of the adult. The shape of the opisthosoma is usually oval or globular, but it is flattened in Trogulidae. It is covered with a fairly hard exoskeleton often decorated with spines. In European species these are small and usually amount to no more than rows of points indicating segments, but in many tropical forms, as in the Gonyleptidae, there are more elaborate developments.

In most Palpatores, nine tergites and nine sternites can be recognized, followed by the operculum anale, a small circular plate pierced by a transverse or round anus. This operculum is regarded as the tergite only of the tenth somite. Hence in these two sub-families there are ten

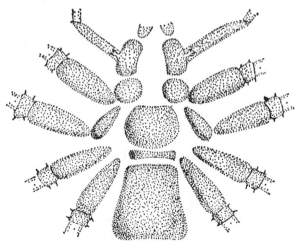

FIG. 69. Homalenotus, ventral aspect, expanded.

opisthosomatic somites, of which the last is incomplete, but in many Palpatores the ninth and sometimes also the eighth somites are much reduced or missing. In Laniatores there are generally nine sternites and eight tergites with the operculum anale.

The general arrangement is that the first five tergites are fused together into a single dorsal shield which is often also fused with the carapace. The last three somites with the anal operculum are free and are usually directed downward. The first sternite surrounds the genital orifice and in general the second sternite is extended forward forming a long plate which carries the genital orifice into a curious position not far behind the mouth.

DISTRIBUTION

Although Opiliones are one of the orders whose distribution may be summarized by the word ubiquitous, there are several features of interest.

The Laniatores (Fig. 70) are the most highly specialized sub-order. They number over 900 species and are the dominant group in southern latitudes. Of the six families, Oncopodidae are a small group confined to South-East Asia, while Cosmetidae and Gonyleptidae are found in South and Central America and include almost a quarter of the sub-order. The Phalangodidae are widespread and are also the only European family of the sub-order. Most of the species are small and generally live in caves. The Triaenonychidae are found only in Australia and Africa, where they are the dominant family in the southern African fauna.

The Palpatores are only slightly less numerous than the Laniatores and include all the familiar Opiliones of the north temperate regions. They extend across the tropics, but are less numerous in hot countries so that they are to this extent supplementary to the other sub-orders.

The Trogulidae are a family of slow-moving species, flattened in form and generally found among moss or grass in damp situations. The dorsal surface is covered with a long shield and the front edge of the prosoma is prolonged into the spinous hood or camarostome characteristic of the family. The Nemastomatidae have a moderately hard epidermis and conspicuously long pedipalpi, usually without a claw. The Ischyropsalidae are a small south European family, most generally found in caves and in mountainous districts. Their peculiar feature is the length of their chelicerae, which in some species are as long as the body. The Acropsopilionidae contain one genus of one species from Chile, and another also of one species from South Africa.

The Phalangiidae include all the commonest harvestmen of Europe and North America, and are of very wide distribution.

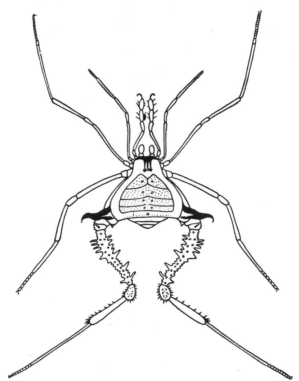

Fig. 70. Laniatores, dorsal aspect. After Berland. Species, *Gonyleptes janthinus*.

PALAEONTOLOGY

Fossil Opiliones are not unknown. From the Carboniferous comes the genus Nemastomoides, with three species, *N. elaveris*, which has been found both at Commenty in France and at Ellismuir in Scotland, and *N. longipes* and *N. depressus*, both from Illinois. The genus Dinopilio is now placed among the spiders.

Seven genera of the family Phalangiidae are represented among the harvestmen found in Baltic amber, as well as two species from the Miocene of Florissant. Among the most interesting is the species *Caddo dentipalpis*. This genus is represented today by only two very rare species, restricted to the eastern United States and Canada. It is characterized by large eyes on a low divided tubercle, and the fossil species is very typical and is closely allied to the living ones.

The sub-order Laniatores has left us only one species, *Gonyleptes nemastomoides*, from Baltic amber.

CLASSIFICATION

The foundations of systematic opilionology were laid by Sørensen, Thorell and Simon, and their methods have never needed revision, but only amplification.

Sørensen in 1873 was the first to divide the order into two families, Gonyleptidae and Opilionidae, and three years later Thorell raised these to the rank of sub-orders, based on the character of the pedipalpi, under the names Laniatores and Palpatores.

Simon in 1879 altered the names of these sub-orders to Mecostethi and Plagiostethi, because his division was based on the character of the sternum. He also added a third sub-order, Cyphophthalmi, which is more closely allied to the Mecostethi than to the Plagiostethi, and which included the family Sironidae, with the fabulous Gibbocellum.

Roewer's great work, "Die Weberknechte der Erde", appeared in 1923 and adopted the same sub-orders, with descriptions of about 1,600 species in 12 families. The first divisions of the order are given in the following table.

SEPARATION OF THE OPILIONES INTO SUB-ORDERS

---

1 (2)	Pedipalpi strong, tarsi with strong reflexed claw; anterior legs with single claw, posterior legs with two or three tarsal claws	LANIATORES
2 (1)	Pedipalpi feeble, with small claw or with none; all legs with a single tarsal claw	PALPATORES (3)
3 (4)	Palpal tarsus shorter than tibia, claw rudimentary or absent	Group DYSPNOI
4 (3)	Palpal tarsus longer than tibia, claw well-developed	Group EUPNOI

Knowledge of the Laniatores has been extended and their systematics modified by Briggs (1969), who has examined species from Oregon and Washington in which the ninth abdominal tergite was free, instead of being either fused to the anal plate or absent. Others were found with three lateral abdominal sclerites. These and other features required the establishing of two new families, the Pentanychilidae and Erebomastridae. His key to the sub-order, slightly abbreviated, is:

SEPARATION OF THE LANIATORES INTO FAMILIES

---

1 (2)	Third and fourth tarsi with two simple claws	3
2 (1)	These tarsi with a complex claw, having a single point of attachment	11
3 (4)	Last two to four abdominal tergites movable, the anterior ones forming a short scute	5

4	(3)	Only last tergite (anal plate) movable, the rest forming a dorsal scute	ONCOPODIDAE
5	(6)	Tarsi 3 and 4 with dorsal projection between the claws (pseudonychium); fourth coxae broad	7
6	(5)	Tarsi 3 and 4 with no pseudonychium; fourth coxae normal	9
7	(8)	Tibia of palpus compressed, with fine setae	COSMETIDAE
8	(7)	Tibia of palpus not compressed, with stout spines	GONYLEPTIDAE
9	(10)	Fore margin of scute with 3, 5 or 7 spines, palpi weakly armed	ASSAMIIDAE
10	(9)	Fore margin of scute without spines, palpi with elongate spines	PHALANGODIDAE
11	(12)	Third and fourth tarsal claws with a median prong	13
12	(11)	These claws bifurcate, with no median prong	EREBOMASTRIDAE
13	(14)	Third and fourth tarsal claws with broad median plate or thickened prong	15
14	(13)	These tarsal claws with continuous median prong	17
15	(16)	Third and fourth tarsal claws with two well-developed branches, approaching length of median prong	SYNTHETONYCHIDAE
16	(15)	These claws with small branches, shorter than the median prong	TRAVUNIIDAE
17	(18)	Ninth tergite distinct from anal plate	PENTANYCHILIDAE
18	(17)	Ninth tergite absent or fused to anal plate	TRIAENONYCHIDAE

SEPARATION OF THE PALPATORES INTO FAMILIES
DYSPNOI

1	(2)	No ocular tubercle, two eyes widely separated	ACROPSOPILIONIDAE
2	(1)	Eyes close together on an ocularium	3
3	(4)	Ocularium on fore-edge of prosoma, forming a hood which covers the mouth parts	TROGULIDAE
4	(3)	Ocularium remote from fore-edge of prosoma	5
5	(6)	First and fourth femora with tubercles or spines	NEMASTOMATIDAE
6	(5)	Femora without tubercles	ISCHYROPSALIDAE

EUPNOI

1	Only one family	PHALANGIIDAE

Simon's system, on which the above tables are founded, has been followed with few changes since 1879. Suggestions and the addition of new families during recent years have foreshadowed a new classification, but the time for this is not yet. The following is an outline of the scheme towards which thoughts seem to be moving.

Order Opiliones
    Sub-order Oncopodomorphae
        Family Oncopodidae

Sub-order Laniatores
  Super-family Gonyleptoidea
    Family Gonyleptidae
        Cosmetidae
        Assamiidae
  Super-family Travunoidea
    Family Travuniidae
        Triaenonychidae
        Erebomastridae
        Pentanychidae
        Phalangodidae
        Synthetonychidae
Sub-order Palpatores
  Super-family Caddoidea
    Family Caddidae
  Super-family Phalangioidea
    Family Phalangiidae
        Leiobunidae
        Neopilionidae
        Sclerosomatidae
  Super-family Troguloidea
    Family Trogulidae
        Nemastomatidae
        Ischyropsalidae
        Sabaconidae

"The study of harvestmen is the study of legs" forms a slogan for all opilionologists, for a harvestman's legs are its main sources of information; they do not replace its eyes, but certainly they supplement them; they are also its nose and its tongue and perhaps also its ears. They are led by the legs of the second pair, which are so indispensable that the loss of one of them is a serious handicap; the loss of both is quickly fatal. In a sense, every order of animals is unique and peculiar to itself, but few are more remarkable or more surprising than are the Opiliones.

The harvestmen are, surely, the comedians among Arachnida: animals with rotund bodies ornamented with little spikes, with two eyes perched atop, back to back, like two faces of a clock-tower, with ungainly legs insecurely attached, with feeble jaws and an undying thirst—a queer assortment of characters, even among a queer folk. Dr Johnson's scornful comment on Gulliver was, "Once you have thought of big men and little men, the rest is easy". Perhaps the same principle applies here, "when once you have thought of long legs and short bodies, the rest is easy". But easy or not, the harvestmen are one of the most strongly individual orders of Arachnida: the biology of the European species has been well studied, but there are certain to be surprises to come from those that live in remoter places.

25

The Order Cyphophthalmi

[Cyphophthalmi Simon, 1879; Anepignathi Thorell, 1882; Cyphophthalmini Petrunkevitch, 1955]

Arachnida in which the prosoma has an unsegmented carapace usually convex posteriorly. Eyes, when present, are two sessile ocelli. The ducts of the odoriferous glands open on short conical processes. The opisthosoma is of ten somites. The chelicerae are chelate, of three segments, the proximal elongated. Pedipalpi of six segments, with a tarsal claw. Legs of seven segments, the tarsus single or of two pieces only, the claw smooth. The sternum is absent. Males are distinguished by the existence of three anal glands, and by a gland in the tarsus of the fourth legs.

The prosoma of the Cyphophthalmi is protected by a uniform carapace, in which no trace of the original segmentation persists. It is usually slightly concave in front and markedly convex behind. It is liberally covered, as is the abdomen, with small pointed tubercles and short spines. Eyes are present only in the family Stylocellidae; they are small and round and are placed just above the coxae of the second pair of legs. They are level with the surface. The orifices of the ducts from the odoriferous glands are found near the apex of two short tubercles, opposite the interval between the second and third coxae. There are a few lyriform organs in the posterior region (Fig. 71).

No sternum is visible in this order.

The opisthosoma clearly shows its segmented nature; ten tergites and nine sternites can be recognized. As seen from above, nine tergites cover the opisthosoma of Siro; on the ventral surface the first three sternites are fused together, the three following are distinguishable but are joined. The seventh sternite lies freely. The eighth and ninth are curved, are narrower in the middle than at the sides, and form the corona analis, closely associated with the diminutive tenth tergite, which acts as a valve to the anus.

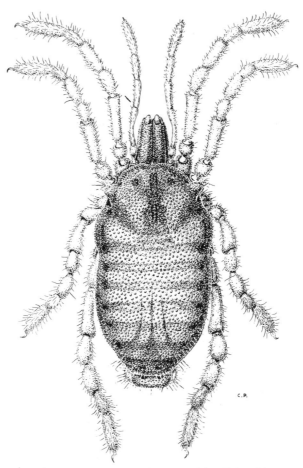

FIG. 71. *Siro rubens*; dorsal aspect. After Juberthie.

The appendages of the adult have their own characteristic features.

The chelicerae retain the primitive number of three podomeres, of which the first or proximal member is cylindrical, points directly forwards and is proportionately longer than in Opiliones. The second part bears a row of conspicuous setae and is prolonged to form the fixed finger with eight teeth. The movable finger articulated to it has seven teeth. The whole is covered with short pointed tubercles and many short setae.

This type of surface is repeated on the pedipalpi. These are of six podomeres, the tarsus a little longer than the tibia and armed with a smooth claw.

The legs are short; in nearly all species they are about the same length as the body. This is peculiar; many Cyphophthalmi live in caves, and most cave-dwelling invertebrates have evolved exaggeratedly long legs. Maxillary lobes are present on the first and second coxae.

One of the characteristic features of the Cyphophthalmi is the presence of a secondary sexual difference on the legs of the males. The fourth tarsus is swollen and carries near its centre and on the dorsal surface a small apophysis, containing a gland, the duct from which opens just below the tip of its basal portion (Fig. 72). The function of this gland and the nature of its secretion appear to be unknown; a pheromone may reasonably be expected.

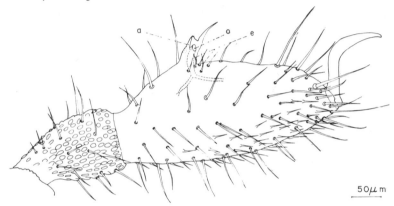

FIG. 72. Tarsus of male Siro. After Juberthie. Abbreviations: (a) apophysis of tarsal gland; (e) short seta at base of apophysis; (o) orifice of tarsal gland.

A second difference between the sexes is found in the existence in the male of three anal glands, not present in the female. The group consists of a pear-shaped median gland and two nearly spherical lateral glands. All contain cells of two types, large and small: their ducts open on the dorsal surface, close to the hind edge of the abdomen. Again, the function of the glands is unknown. Their position recalls that of the gland in the false scorpion Serianus, where they assist in the formation of the spermatophore. Should these be homologous, further support would be given to the belief that the Cyphophthalmi also make use of a spermatophore, an important difference between these Arachnida and their closest relatives.

The most significant feature of the genitalia is the nature of the male organ (Fig. 73). Unlike the long, chitinous penis of Opiliones, that of the Cyphophthalmi is short, thick and but feebly chitinized. It is in two segments, of which the distal part carries several long setae. The ovipositor, on the other hand, is almost as long as the female's body. It

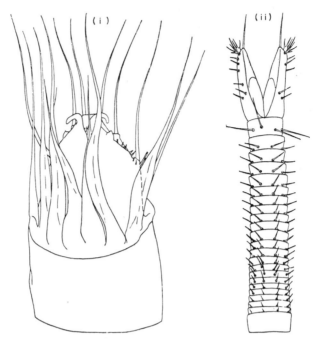

FIG. 73. Cyphophthalmi: (i) penis and (ii) ovipostor. After Juberthie.

consists of about 30 rings and ends in a pair of long lobes, the whole reminiscent of the ovipositor in the Phalangodidae.

Most probably it is used in the same way. In groups of three to five, eggs are laid in crevices in the soil at intervals from spring to autumn: they are small green spheres, less than half a millimetre in diameter. Within a week a larva is hatched; it cannot feed, but lives on the yolk in its gut until, hiding itself in some safe spot, it moults. Figure 74 shows how different is this larva from any early stage of any other arachnid, including even the Opiliones. Ecdysis changes the larva into a nymph, resembling the adult, save in size. Its behaviour between one ecdysis and the next is characteristic. For seven or eight months it lives a normal life in the cryptosphere; it then digs a hole into which it shuts itself with the loosened soil. Here it remains for four months or so, until it emerges a large nymph, a year older than the nymph that dug the hole. Four years may be occupied in this remarkably slow development before a fourth ecdysis produces an adult. Adult life may last for five years, giving a total life span of nine years, a period of time scarcely approached by any other small arachnid.

The food taken during this time seems to consist of Collembola and Protura, animals that supply the usual diet of most of the Cryptozoa.

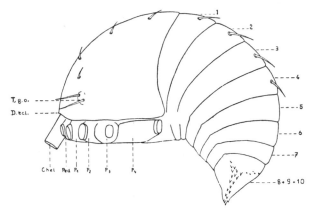

FIG. 74. Larva of Siro. After Juberthie.

A cyphophthalmid can protect itself from aggressors by the use of its malodorous secretion. Threat causes the appearance of a little liquid at the orifice of the duct, one of the legs is made to touch this, and the drop deposited on the aggressor. This direct method of bringing the repellant to the foe is peculiar to the order.

DISTRIBUTION

The first species of the order, *Siro rubens*, was discovered by Latreille at Brive in the south of France in 1796, and it is from France that much of our recent knowledge of these Arachnida has come, and which we owe to the work of C. Juberthie. The order is widespread in the tropical and temperate zones, where a proportion of the species are found only in caves. As the map (Fig. 75) shows, their occurrence, as so far known, is discontinuous, a feature characteristic of many primitive animals. Their range stretches from Oregon to New Zealand, from Chile to Japan.

CLASSIFICATION

Two families have been distinguished, Sironidae and Stylocellidae. The former includes some 15 genera, the latter about ten. The typical genus, is also the best known, the genus Rakaia, with 20 species, mostly from New Zealand, and the largest. The chief difference between the families is the existence of eyes in the Styllocellidae while the Sironidae are blind.

The removal of the Cyphophthalmi from the Opiliones, with the consequent establishing of a new order, was not undertaken without considerable thought and correspondence with other arachnologists (Savory, in press).

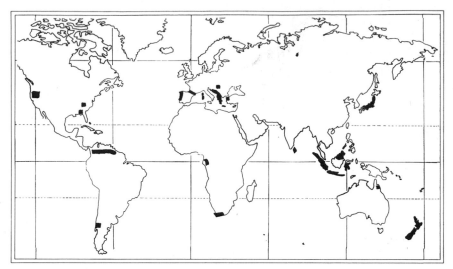

Fig. 75. Map showing the distribution of Cyphophthalmi.

The resemblances between the two taxa were close enough to unite them when the Opiliones were less well known and the Cyphophthalmi were scarcely known at all. Their separation follows fuller knowledge of their structure and behaviour, and this, as always, has tended to emphasize distinctions rather than similarities. It is common taxonomic experience that the splitting of any group is a consequence of meticulous study and a recognition of the significance of differences.

The most obvious similarity between Cyphophthalmi and Opiliones is the presence of odoriferous glands in the prosoma. While it is true that all Opiliones possess odoriferous glands, it is not in logic true that all possessors of such glands are Opiliones.

Another resemblance is seen in the form of the ovipositor, so that there is an undeniable relation between the two orders, with the addition that the Cyphophthalmi resemble the Laniatores—and especially the Oncopodidae—more closely than they resemble the Palpatores, while remembering that they also resemble the Notostigmata among the Acari.

The differences between the Cyphophthalmi and the Opiliones may be summarized in short form thus:

(1) The sculpturing of the exoskeleton
(2) The nature of the bodily segmentation
(3) The position of the eye (when present)
(4) The form of the penis

(5) The use of a spermatophore
(6) The absence of a genital operculum
(7) The tubercles for the odoriferous glands
(8) The tarsal glands of the males
(9) The anal glands of the males
(10) The length of life

It is maintained that the above characteristics are sufficient in number, in detail and in principle to justify the creation of the order. This is a step of which some will undoubtedly disapprove, and to the strength of their disapproval the following concession may be offered as a general conclusion.

From an ancestral group of proto-Opiliones there evolved the primitive cryptozoic order that has been described. Either from the same ancestry or from some early Cyphophthalmi there evolved the somewhat less photonegative, less thigmotropic order which increased in dimensions, forsook the cryptosphere, abandoned the spermatophore and acquired the use of a true intromittent male organ. There is sufficient suggestion that a similar course of evolution has occurred among other sets of related arachnid orders to support this interpretation of the phylogeny of the Cyphophthalmi.

In an elementary classification, based on similarities of external appearance, the Cyphophthalmi may well be left to obscurity among the Opiliones; but in a phylogenetic classification, which attempts to reflect the course of evolution, the distinction between the free-living and specialized Opiliones and their primitive, cryptozoic forerunners is too conspicuous to be overlooked. It would be irrational to do so.

26

The Order Acari

[Monomerostomata Leach, 1815; Acari Sundevall, 1833; Acarina Nitzsch, 1818; Acarides Heyden, 1826; Acarenses Dugès, 1834; Acarida Englemann, 1860; Rhyncostomi]
(Mites, Ticks)

Arachnida, often highly specialized, in which modifications of the segmentation divide the body into a proterosoma and a hysterosoma, usually distinguishable as a boundary between the second and third pairs of legs. There are often no eyes. The chelicerae are of three segments, chelate or piercing or otherwise modified. The pedipalpi are of six segments, variously modifiable. The legs are very variable, the typical number, eight, being sometimes replaced by six, four or two, and with the segments, often six in number, varying from two to seven. The respiratory apertures and genital openings vary considerably in position. Predatory, phytophagous and parasitic forms occur.

Because the Acari differ from most other arachnids in their inconspicuous somatic segmentation and because of the variety of modification of body structures it is inconvenient to retain the normal simple subdivision of the body into prosoma and opisthosoma. Instead the following nomenclature is used to classify the body divisions (see Fig. 76).

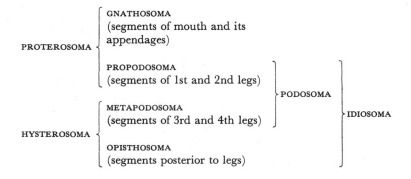

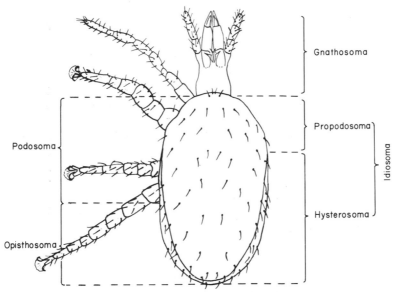

Gnathosoma

Propodosoma

Idiosoma

Hysterosoma

Podosoma

Opisthosoma

FIG. 76. Divisions of the body of a mite.

Yet another way in which the Acari differ from other Arachnida is in their life cycles. The majority of mites are oviparous although some mesostigmatid and prostigmatid mites exhibit ovoviviparity, and viviparity has been reported in some parasitic forms (Mesostigmata). The eggs, or ova, are ovoid and may be deposited either singly or in batches. In some prostigmatid mites as few as four eggs may be laid *in toto* whilst some members of the Metastigmata may produce as many as 10,000.

The mature embryo usually hatches by rupturing the eggshell but in many cases a second membrane, called the deutovial membrane, is formed inside the outer coat. Where this occurs it expands when the hard outer coat ruptures and the emerging form is called a pre-larva or deutovum. The removal or partial removal of the hard outer coat allows growth to occur before actual hatching. If the removal is incomplete then the presence of the membrane affords some protection in the developing deutovum. The Prostigmata and Cryptostigmata are examples of groups of mites in which deutova may be found.

With but few exceptions, the newly hatched stage, called the larva, has six legs. This eventually moults to become an eight-legged nymph which closely resembles the adult in some species whilst in others it differs markedly. Apart from the ticks and free-living mesostigmatid mites there are usually three nymphal stages, each one being progressively larger than the previous. These three nymphal stages are called the protonymph, the deutonymph and the tritonymph. In one of the groups of ticks, commonly called the hard ticks, there is only one nymphal stage whereas in the other group, the soft ticks, there may be as many as eight. In the free-living mesostigmatid mites there are usually two nymphal stages (Fig. 77).

In many astigmatid mites, the deutonymph is a resistant stage and differs from the preceding and succeeding nymphs in morphology and behaviour. It is called the hypopus and is formed when conditions are unfavourable. A hypopus does not occur in the life cycles of the skin parasites (Sarcoptidae and Psoroptidae) which pass through only one larval and two nymphal stages before becoming adult.

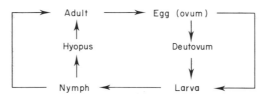

Fig. 77. The life cycle of the Acari.

Although Acari show a wide variety of adaptations regarding their feeding habits, the mouth parts may be relegated to a common plan which shows similarities with those of other Arachnida. The mouth parts (chelicerae and pedipalpi) are borne on the anterior gnathosoma which is little more than a hollow tube leading to the mouth. The roof of this tube is the tectum capituli, the lateral walls are the fused pedipalpal coxae and the floor is composed of the subcapitulum and the endites of the pedipalpal coxae which are prolongated to form the hypostome. This is especially conspicuous in ticks (Fig. 78).

The chelicerae together with the pedipalpi are the food-acquiring organs. The chelicerae are variously modified, being chelate in some scavenging and predatory forms where they are used for prehension, or stylet-like, or armed with movable cutting digits as in some parasitic forms. Variation is seen in the pedipalpi also and these may be leg-like, or large and clawed, or reduced in size and used as sense organs. In certain fur mites of the Astigmata there are two lobe-like expansions of the pedipalpal coxae which form clasping organs.

Acari are fluid-feeders and either feed on liquid food or render their meal fluid by secreting enzymes from the salivary glands. Once liquefied the meal is imbibed by the action of the suctorial pharynx.

Owing to the large number of species of Acari which exist and because of their great diversity, it is more convenient to deal with the character-

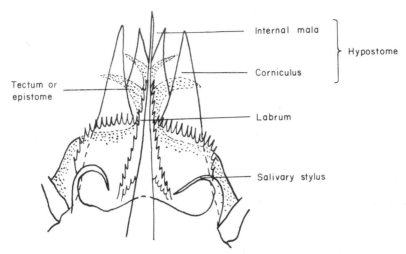

FIG. 78. Gnathosoma of a tick, *Ixodes ricinus*.

istics of each sub-order in turn. In the classification used here the order Acari is divided into seven sub-orders. The distinction between these taxa and some details of the biology of the members are given below.

Notostigmata

Members of this sub-order are large mites (1,000 μm+), elongated, unsclerotized and leathery in texture. Their legs are long and slender and they resemble harvestmen superficially. They have four pairs of small dorso-lateral spiracular openings, located on the hysterosoma at a level behind the fourth coxa. Notostigmatid mites show a preference for dark, semi-arid environments and may be found under stones, rocks and organic debris in parts of North America, South America, Central Asia and the Mediterranean. These mites are omnivorous, feeding on small arthropods and pollen grains. There is only one family, the Opilioacaridae, containing a few genera such as Opilioacarus.

Tetrastigmata

Mites in this sub-order are large (2,000–7,000 μm), heavily sclerotized and non-segmented externally. They possess a pair of ventro-lateral spiracular openings at the level of the third coxae. A second pair of openings located behind the fourth coxae, and called the air sac pores, may be homologous with the expulsory vesicles found in some free-living Mesostigmata.

Tetrastigmatid mites are predatory and are found in Australia, New Zealand, New Guinea, Seychelles, Mauritius and Sri Lanka. The order contains a single family, the Holothyridae, usually regarded as having a single genus Holothyrus.

Mesostigmata

With the Mesostigmata we come to a large and successful group of mites which have radiated to occupy a variety of habitats. Although the majority of members of this sub-order are predators, some are scavengers and many are external or internal parasites of mammals, birds, reptiles or invertebrates. The size range of these mites varies from 200 to 2,000 μm or more, and they are characterized by the possession of a number of sclerotized plates on the dorsal and ventral surfaces and a single pair of spiracular openings between the second and fourth coxae (Fig. 79).

These mites are distributed throughout the world and individual species have become adapted to life in the soil, leaf-litter, animal nests,

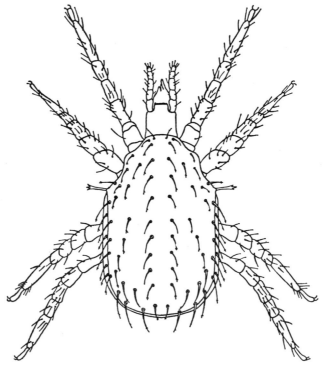

FIG. 79. Mesostigmata. *Zercoseius ometes.* After von Vitzthum.

stored food, and in and on animals and plants. The sub-order is divided into about 60 families.

An example of a mesostigmatid mite leading an ectoparasitic mode of life is *Dermanyssus gallinae* the red poultry mite, which feeds on the skin surface. It hides during the daytime and attacks the birds at night. This species also causes dermatitis in humans. Other mites have invaded body apertures such as the tracheae, air-sacs, bronchi and lung tissues as in *Sternostoma trachaecolum* of canaries and *Ptilonyssus hirsti* of the house sparrow. *Ophionyssus natricus* is a very common ectoparasite of snakes: feeding stages are usually found in the eye sockets and beneath the scales of the head. They transmit *Pseudomonas hydrophilus*, a flagellate which causes severe and often fatal haemorrhagic septicaemia in reptiles.

Many mesostigmatids are free-living and predacious. Examples of such mites are the species of the genus Zercon. The mite *Blattisocius tarsalis* is interesting in that it is found in stored grain and feeds, not on the stored food itself, but on other grain-feeding mites.

Metastigmata

Members of this group of mites differ so radically in their size from their relatives that they have been given a separate common name and are known as "ticks". Adults range from 2 to 6 mm in length when unfed and females of the larger species may reach over 2 cm in length when fully engorged.

Ticks are ectoparasitic in all of their post-embryonic stages and feed on the blood and tissue fluids of mammals, reptiles and birds. An exception to this feeding pattern occurs in *Aponomma ecinctum* which parasitizes a beetle. The hypostome of ticks is equipped with recurved teeth and is used, not as an aid to skin penetration, but as a holdfast organ. Penetration is effected by means of the chelicerae and enzymes from the salivary glands.

Characteristically ticks have a sensory organ on the tarsus of the first leg called Haller's organ which is equipped with olfactory and hygro-receptor setae, and a single pair of spiracular openings close to the coxae of the fourth legs. Larval ticks lack such respiratory openings.

There are two main families of ticks: the Ixodidae or hard ticks, possessing a hard dorsal shield or scutum, with 700 species in nine–12 genera, and the Argasidae or soft ticks which lack such a shield and which are organized into four genera. The hard ticks have three feeding stages in their life cycle (larva, nymph and adult) and normally each feeds on a different host animal. An example of such a "three-host" tick is *Ixodes ricinus*, the British sheep tick (Figs 80 and 81). Modifications to this pattern occur, however, with some species such as *Rhipicephalus*

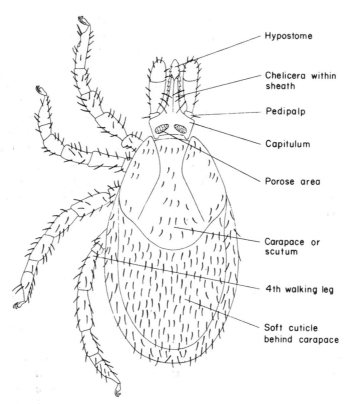

Hypostome

Chelicera within
sheath

Pedipalp

Capitulum

Porose area

Carapace or
scutum

4th walking leg

Soft cuticle
behind carapace

FIG. 80. Metastigmata. *Ixodes ricinus*; dorsal aspect. After Snow.

evertsi using only two hosts and *Boophilus annulatus* parasitizing only one. In the former case the larva and nymph feed on the same animal and in the latter case all three stages utilize the same host. Another modification regarding feeding is that in many species the adult males do not feed and in those which do they imbibe only small quantities of food. This situation compares dramatically with the females which may suck over 50 times their own weight of blood and other fluids, taking several weeks to complete their single meal. Whereas the hard ticks are found mainly on mammals, the Argasidae are most commonly, although certainly not exclusively, located on birds. They feed intermittently, often on the same animal and keep in the nesting area of the host. Most species are parasitic in all stages, e.g. *Argas persicus*, while some feed only as larvae and nymphs (*Otobius megnini*) and others in all but the larval stage (*Ornithodoros savignyi*).

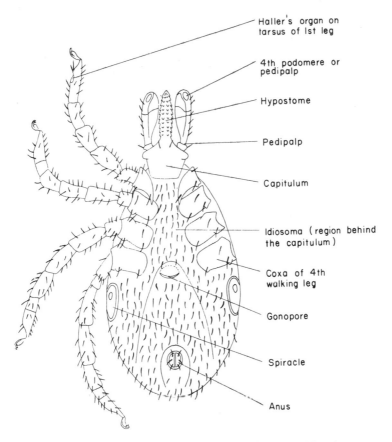

Haller's organ on tarsus of Ist leg

4th podomere or pedipalp

Hypostome

Pedipalp

Capitulum

Idiosoma (region behind the capitulum)

Coxa of 4th walking leg

Gonopore

Spiracle

Anus

FIG. 81. Metastigmata. *Ixodes ricinus*; ventral aspect. After Snow.

Many ticks are important pathogen transmitters disseminating such important diseases as red-water fever, forms of encephalitis, relapsing fever and tularaemia.

Prostigmata

It is generally recognized that this group is a heterogeneous assemblage of mites which share certain common features. Amongst these characters are that the spiracular openings, when present, are paired and are located either between the chelicerae or on the dorsal surface of the propodosoma. Prostigmatid mites are usually weakly sclerotized and have chelicerae which vary from being strongly chelate to reduced, and pedipalps which may be simple, fang-like or clawed. (Fig. 82).

The group contains terrestrial, aquatic and marine forms and includes

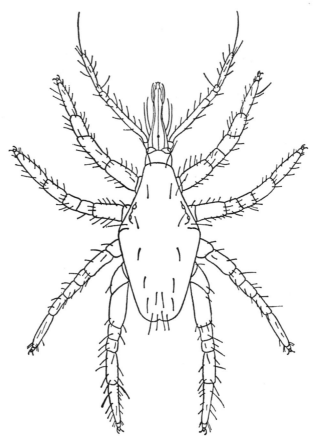

FIG. 82. Prostigmata. *Scirus longirostris*. After von Vitzthum.

predators, phytophages and parasites. Smaller members are only 100 or
so μm in length whilst the upper limit of length is in the order of 1·5 cm.
One-hundred-and-thirteen families are usually recognized in this
sub-order.

Members of the Prostigmata are cosmopolitan in their distribution,
and their habitats are almost unlimited. The group contains the aquatic
and semi-aquatic water mites which are small, round-bodied, brightly
coloured inhabitants of fresh and salt water. The larvae of water mites are
usually parasitic on other arthropods whereas the later stages are predacious. The gall and rust mites such as *Aculus cornutus* found on peach and
Aceria fraxinivorus on ash are also members of the Prostigmata. Several
gall mites have been demonstrated to be the transmitters of plant virus
diseases. Other members of the group include the snout mites or Bdel-
lidae which are active red, brown or green mites preying on small arthro-

pods and arthropod eggs, and the colourful spider mites some of which seriously damage orchard trees and food crops, for example *Tetranychus cinnabarinus*, a pest of cotton. In addition to these free-living and plant-parasitic species there are many which parasitize animals. Among them are the trombiculids such as *Trombicula autumnalis* (Fig. 83), the European harvest mite, which may cause dermatitis in man. Species of Trombicula are parasitic in the larval stage only, the nymphs and adults being free-living predators. In the Oriental and Australasian regions Trombicula species transmit *Rickettsia tsutsugamushi* which causes scrub typhus.

Astigmata

With this cosmopolitan group of mites there is a departure from a predacious existence. Instead many members are fungivores and

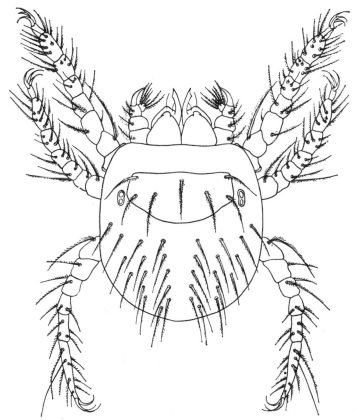

FIG. 83. Prostigmata. *Trombicula autumnalis*, larva. After von Vitzthum.

saprophages whilst others are graminivores and some parasitic. Amongst the parasites are many skin and feather mites of birds, skin parasites of mammals and respiratory and visceral parasites of vertebrates.

Astigmatid mites are mostly slow moving and weakly sclerotized. They range in length from 200 to 1,500 μm and respire through their integument. The group is divided into some 40 families.

An important astigmatid mite is *Acarus siro* which feeds directly on stored grain. It is of economic significance since it destroys the wheat germ. Of the parasitic mites the ear mite of dogs and cats, *Otodectes cynotis*, is important as is the feather mite of domestic fowl and some other birds, *Megninia cubitalis*. *Sarcoptes scabiei* is a mite which attacks the skin of man as well as a number of other animals. It causes a condition known as scabies or sarcoptic mange. Other mites which affect man are the house dust mite, *Dermatophagoides pteronyssinus*, which has been implicated as the cause of house dust allergies, and *Dermatophagoides scheremetewskyi* which causes eczema seborrhica.

Cryptostigmata

Commonly known as oribatid mites or beetle mites, because of their superficial resemblance to these insects, this is a cosmopolitan group of heavily sclerotized mites which range in length from 200 to 1,500 μm. Members of the 100 or so families of cryptostigmatids are mainly fungivorous, saprophagous or algivorous and are common in soil, leaf-litter and under stones and bark.

Spiracles are absent in these mites although some families possess tracheae associated with paired dorsal pseudostigmata and with the bases of the first and third legs.

Species of the genus Oppia are common inhabitants of forest humus and grasslands. A common example is *Oppia ornata* found in mixed oak and beech humus. Another species of this genus, *O. minuta*, is an important vector of the sheep tapeworm *Moniezia expansa*, as too is *Galumna virginiensis*.

DISTRIBUTION

The Acari are ubiquitous, for their small size, their modes of life and their ability to withstand adverse conditions all help them to achieve universal distribution. The most northerly land in the world, the north coast of Greenland, numbers mites among its fauna and they have also been found not only on the fringe of the Antarctic, but also on the Shackleton Glacier in 86° 22′S. Mites attain great altitudes, and are, for example, plentiful in the fleece of the sheep which live at a height of above 4,500 m on the plateaux of Turkestan. At much greater heights than this they are to be found drifting in the aerial plankton.

PALAEONTOLOGY

The oldest known fossil mite belongs to the Devonian age, and was found in the Old Red Sandstone of Aberdeen. Its name is *Protacarus crani*, and to some extent it resembles the recent Eupodoidea of the Prostigmata, but has also some primitive characters.

Oligocene amber contains a fair number of isolated genera. Some of these appear to be extinct, but some, especially among those belonging to Cryptostigmata, belong to recent genera, and cannot actually be distinguished from living species.

From peat, galls have been obtained some of which may well have been caused by members of the Prostigmata.

CLASSIFICATION

The classification of the Acari has passed through the stages common to all large groups of animals, the known species of which are constantly increasing in number. At present nearly 2,000 genera and some 20,000 species have been described and it is estimated that up to 500,000 more species may exist.

A brief look at some earlier systems of classification is useful as it helps in the understanding of present schemes.

The system proposed by Banks in 1905 was in general used for over 20 years, and for its simplicity has much to recommend it. He divided the *order* Acari into eight super-families as follows:

(1) Eupodoidea	(5) Gamasoidea
(2) Trombididoidea	(6) Oribatoidea
(3) Hydrachnoidea	(7) Sarcoptoidea
(4) Ixodoidea	(8) Demodicoidea

Oudemans in 1906 forwarded a different classificatory system based mainly upon the number of respiratory plates (spiracles or stigmata). He divided the *class* Acari into five sub-classes:

(1) Astigmata
(2) Lipostigmata
(3) Zemiostigmata
(4) Octostigmata
(5) Distigmata

Later, in 1923 Oudemans proposed a new system in which the then *order* Acari was divided into six sub-orders:

(1) Notostigmata
(2) Holothyroidea
(3) Parasitiformes
(4) Trombidiformes
(5) Sarcoptiformes
(6) Tetrapodili

Vitzthum (1931 and 1940-43) generally recognized Oudemans' (1923) classification, but Grandjean (1935) made a division of the Acari based on the presence or absence of actinochitin in the setae. This same character has been used by Evans, Sheals and Macfarlane as recently as 1961 who divided the Acari into two super-orders, the Acari-Anactinochaeta and the Acari-Actinochaeta. This character is not used by Krantz (1970) although his order Acariformes is equivalent to the Acari-Actiniochaeta of the previous authors and Krantz's Opilioacariformes and Parasitiformes together equate with Evans, Sheals and Macfarlane's Acari-Anactinochaeta:

KRANTZ	GRANDJEAN
Order Opilioacariformes	Super-order Acari-Anactinochaeta
Sub-order Notostigmata	Order Notostigmata
	Order Tetrastigmata
Order Parasitiformes	
Sub-order Tetrastigmata	
Sub-order Mesostigmata	Order Mesostigmata
Sub-order Metastigmata	Order Metastigmata
Order Acariformes	Super-order Acari-Actinochaeta
Sub-order Prostigmata	Order Prostigmata
Sub-order Astigmata	Order Astigmata
Sub-order Cryptostigmata	Order Cryptostigmata

In almost every animal taxon large or small, there is to be found one group which in range and in numbers far surpasses the rest. The beetles among the insects, the ungulates among the mammals are examples of this, and among the Arachnida, the spiders and mites show the same kind of supremacy.

The mites are the specialists among the Arachnida. In their often parasitic mode of life, in their occasionally vegetarian diets, in their all too frequent harbouring of protozoans and other organisms pathogenic to man and other animals, and even in their attempts to return to the water, they are an individualized group, separate, distinct from the others of their class. They appear not to be as rich in species as spiders, but this inequality is annually disappearing as mites attract more and more attention; and like all specialists their successes are limited by the fact of their specialization. Yet within these limits it must be admitted that they reign supreme.

27

The Order Ricinulei

[Podogonata Cook, 1800; Meridiogastra Karsch, 1892; Ricinulei Thorell, 1897; Rhinogastra Cook, 1899; Ricinuleida Petrunkevitch 1945]

Arachnida in which the prosoma is covered by an unsegmented carapace, with a cucullus joined to its fore-edge. There are no eyes. The opisthosoma is of nine somites, the true first somite is missing and the second forms a pedicel; the last three are reduced to a narrow pygidium. The three median tergites are divided by two longitudinal furrows into three plates each. The sternum is small, and is hidden by the coxae. The chelicerae are of two segments, and are chelate. The pedipalpi are of six segments, chelate, and their coxae are fused in the middle line, forming a camarostome. The legs are of seven segments, with the trochanter and tarsi of the last three variously sub-segmented. All tarsi have two smooth claws. The metatarsus and tarsus of each third leg of the male are modified to form accessory sexual organs.

The prosoma of Ricinulei is protected by a uniform carapace, roughly square-shaped and unusually thick and hard. It is covered, as is the rest of the body, by characteristic granulations, interspersed with setae. Some of these setae are spatulate or club-shaped, with a longitudinal groove and roughened surface: they are probably sense organs of a special kind. The carapace has a central transverse groove of varying length and often a number of other markings, probably indications of some of the original somites. One of the chief characteristics of the order is the cucullus, a wide, oval, slightly convex plate, articulating with the anterior edge of the prosoma. This cucullus is readily movable and when bent downwards it completely covers and protects the mouth and the chelicerae. A similar cucullus is found in some other orders, where it bears the median eyes: the cucullus is therefore regarded as being the first somite of the body, one which is not as a rule separated from those behind it and to which the direct eyes belong. The Ricinulei have no eyes.

Posteriorly the prosoma seems to end in a transverse ridge, but actually the end is below this ridge. This appearance is due to a second characteristic of Ricinulei, the remarkable way in which the prosoma and opisthosoma are linked or clasped together. On the dorsal surface of the opisthosoma there is a deep transverse groove between the second and third tergites, into which the prosomatic ridge fits. Ventrally the third sternite projects forwards and forms a pair of pocket-like spaces into which fit processes from the posterior borders of the fourth coxae. The consequence of this curious arrangement is that the animal does not appear to be pediculate (Fig. 84), the prosoma and opisthosoma are

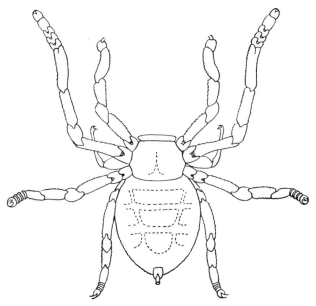

Fig. 84. Ricinulei; dorsal aspect. Species, *Ricinoides crassipalpe*. After Hansen and Sørensen.

securely locked together, and the first two opisthosomatic somites are hidden. Coupling and uncoupling must be possible to the living animal, for the genital aperture lies within the enclosed space; hence during copulation and during egg-laying the attachment must be undone in order to expose the orifice.

The opisthosoma presents the appearance of four well-defined somites, but actually nine can with certainty be distinguished. The first of these is the pedicel. This pedicel has a short narrow tergite, comparatively feebly chitinized, and surrounded by a quite soft and flexible membrane. Its sternite is crescent-shaped, its concave margin facing backward. The second somite is about three times as wide as the pedicel. Its tergite is a narrow strip, its sternite also crescentic, but with its concavity forward. Thus there is between the first two sternites an oval area of membrane in which is the genital orifice, a broad transverse slit. It will be noticed that this places the genital orifice in front of the second opisthosomatic somite, whereas in those Arachnida most closely allied to the Ricinulei it lies on the second somite or on its posterior margin. It may be, therefore, that an undetected somite exists or existed in front of the pedicel, or, alternatively, that the pedicel of Ricinulei is not homologous with that of Araneae or other Caulogastra.

The third somite is short and wide, being a strip of chitin across the opisthosoma where this is coupled to the prosoma. Its tergite is generally divided into a central and two lateral pieces; its sternite is plainly seen on the lower surface.

The fourth, fifth and sixth somites are the largest of the series and constitute the bulk of the opisthosoma. The tergite of each is divided into a large median and smaller lateral areas, separated by softer, lighter-coloured membrane of great thickness. As the animals grow older these passages between the tergal elements tend to narrow and disappear. The sternites of these three somites are not divided longitudinally (Fig. 85), and, as the spaces between them narrow, they come to form an almost continuous shield on the ventral side of the body. Both tergites and sternites of this region are marked each with a pair of depressions, similar to those often seen on the backs of spiders, and like them due to the insertion of muscles within.

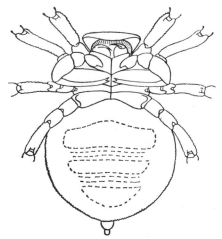

Fig. 85. Ricinulei; ventral aspect. Species, *Ricinoides crassipalpe*. After Hansen and Sørensen.

The seventh, eighth and ninth somites are much reduced in size. Their exoskeletal supports are complete rings of chitin and the segments can be drawn together like a telescope, making them even more inconspicuous. In this condition, which is probably their normal one, the eighth and ninth somites can only be seen inside the seventh, their posterior edges forming concentric circles. The anus is a transverse slit on the last somite: no glands, like those of Uropygi, open beside it.

The sternum is not normally visible from the outside, since it is covered by the pedal coxae. If these are removed it is seen as a small plate, lying longitudinally.

The chelicerae are composed of two segments, and are chelate (Fig. 86). The first, basal, segment is short and stout and provided with a transverse belt of close setae on the ventral surface and a similar smaller belt on its dorsal surface. One or two strong processes, with sharp edges and teeth, rise from the distal end of this segment, and against them works the second segment. This is a sickle-shaped point, often with a serrated edge, and closely resembles the corresponding organ in Araneae.

The pedipalpi are of six segments. They are remarkable in that their coxal segments are fused together in the middle line instead of acting independently as gnathites. In this Ricinulei resemble Uropygi. This common maxillary plate bears the ordinary pair of palpi, consisting of two trochanters, femur, tibia and tarsus. The first trochanter moves slightly up and down, but the second is capable of a complete rotation

FIG. 86. Chelicera of *Ricinoides crassipalpe*. After Hansen and Sørensen.

(Fig. 87), turning through 180° and directing the remaining segments
in the opposite way from that in which they usually lie. The femora are
stout, the tibiae long. The joint between these two allows them to touch
each other when fully flexed and, on account of this mobility, the tip of
the limb can reach the mouth from either direction. The distal end of
the tibia carries a small serrate process against which the tarsus moves
and so makes the pedipalpi also chelate on a small scale.

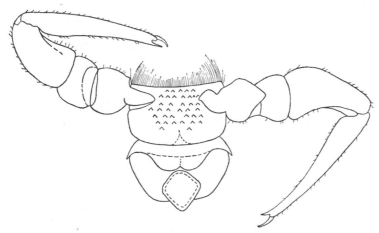

FIG. 87. Pedipalpi of *Ricinoides westermanni*, showing range of rotation. After Hansen
and Sørensen.

The legs of these curious Arachnida also have peculiar characteristics. The first three pairs of coxae are immovably coalesced and the
fourth pair which is concerned with the linkage of the prosoma and
opisthosoma is freely movable. The legs are not all similarly constituted.
The first has one trochanter and one tarsal segment, the second one
trochanter and five tarsals, the third two trochanters and four tarsals
and the fourth two trochanters and five tarsals. The other segments are
as usual, and hence the total number of segments on the four legs are
seven, 11, 11 and 12. The second pair of legs is always the longest, then
the fourth and the first is always the shortest. All the legs are devoid of
ordinary spines and all end in two simple claws without teeth, situated
in a small excavation at the end of the tarsus.

The male organs of Ricinulei are to be found in a unique situation,
on the metatarsus and tarsus of the third pair of legs. The metatarsus
and first two segments of the tarsus are strangely modified by cavities
and by fixed and movable processes. When the tarsus is bent upwards,

it nearly touches the metatarsus and a closed or almost closed space is formed by the interlocking of their processes and cavities.

The female genitalia are shown in Fig. 88. Their disposition suggests that the seventh bodily segment has been repressed and the first sternite belongs to the eighth somite. Pollock (1967) records two observations of mating. The male climbs upon the female's back, his fourth legs grasping her opisthosoma. Insemination is effected by the third legs: there seems to be no courtship nor, apparently, any spermatophore.

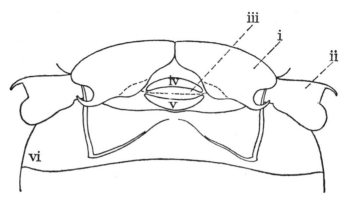

FIG. 88. *Ricinoides afzelii:* female genitalia. (i) Fourth coxa; (ii) fourth trochanter; (iii) vulva; (iv–vi) first, second and third sternites.

In addition to having no eyes, Ricinulei seem to be poorly supplied with sense organs. In the only specimen dissected by Hansen and Sørensen, an immature *Ricinoides crassipalpe,* no lyriform organs were found on pedipalpi, legs or opisthosoma. Nor do Ricinulei possess the long thin spines often described as acoustic setae such as are found on the legs of the Araneae. Their spatulate setae are possibly organs of touch, and a very curious seta is found standing in a small depression on the last tarsal joint of each of the three posterior legs. It is short, slightly broader at its upper than at it lower end and carries delicate scattered branches all over its upper portion. The actual tip is, however, bare.

The tracheal tubes by which Ricinulei breathe open on the prosoma, at two apertures near its apparent posterior margin, above the third coxae. The apertures are very small and are not visible in the intact animal.

DISTRIBUTION

The Ricinulei are of limited distribution, and have so far been found only in scattered regions in the tropical belt (Fig. 89). These are on the

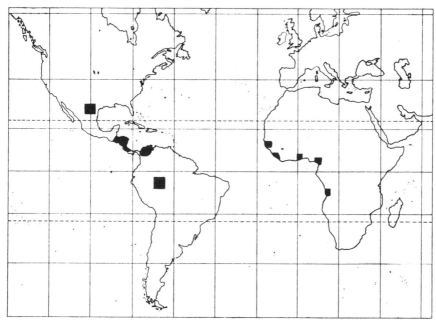

FIG. 89. Map showing distribution of Ricinulei.

west coast of Africa, in Central America, in Mexico and Texas and in the Amazon basin. The two hemispheres contain different genera, Cryptocellus from the New World and Ricinoides from the Old.

PALAEONTOLOGY

This order appears to have been more numerous in Carboniferous times than it is today, for the number of fossil specimens that have been discovered is almost comparable to the number captured as living animals. Nine fossil species have been described, belonging to two genera.

The genus Polyochera contains three species, *P. alticeps*, *P. punctulata* and *P. glabra*. The first of these came from Coseley in Worcestershire, the other two from Mazon Creek, Illinois.

The genus Curculioides dates from 1837, the year before the description of the first living Ricinuleid. It now contains six species, three from America and three from Britain.

These two genera differ from the recent ones in the form of the second coxae. In the living Ricinulei these coxae are broad and meet in the middle line; in the fossil genera they are triangular and do not touch each other. The genus Curculioides was at first placed in a separate

family, Holotergidae, because the opisthosoma seemed to be covered
with a hard unsegmented shield, showing none of the transverse sutures
characteristic of all the other genera.

CLASSIFICATION

The order has at all times been limited to a single family, at first called
Cryptostemmatoidae or Poliocheridae. The name Cryptostemma
Guérin, 1838, for the type genus is, however, preoccupied by Cryptostemma Herrich-Schäffer, 1835, for the Hemiptera, and for this reason
Ewing (1929) proposed the name Ricinoides for the genus and Rici-
noidae for the family. Ewing gives the following table for the classification.

CLASSIFICATION OF THE ORDER RICINULEI

1 (2)	Second coxae subtriangular and not meeting on middle line	3
2 (1)	Second coxae broad, plate-like and meeting on middle line	5
3 (4)	Opisthosoma divided into tergites: palaeozoic	Poliochera Scudder
4 (3)	Opisthosoma uniform: palaeozoic	Curculioides Buckland
5 (6)	Last segment of chelicerae opposed by two processes on penultimate segment: fourth segment of tarsus 2 longer than fifth	Ricinoides Ewing
6 (5)	Last segment of chelicerae opposed by one process on penultimate segment: fourth segment of tarsus 2 shorter than fifth	Cryptocellus Westwood

---

One of the noteworthy features about Ricinulei is due to an accident
which caused someone somewhere to describe them as "primitive
Arachnida", and this adjective was universally applied to them by every
writer until 1945b, when Millot pointed out how inappropriate it is.

In truth the word primitive seems to impose upon its readers a
passive receptivity which prevents them from ever questioning its
accuracy. This is perhaps because the opinion that a given animal is
primitive gives one a comfortable feeling of security. This at least, one
feels, is where we begin. Where do we go from here? The answer to
which question is that we should try to determine in how many ways the
so-called primitive animal has specializations peculiar to itself.

In fact, Mitchell (1972) lists 11 features characteristic of the Ricinulei and distinguishing them from all other orders. So large a number
argues an early origin and a long period of evolutionary specialization.

The second remarkable fact about Ricinulei is their rarity. The first
living specimen was found in 1838, and by 1931, 93 years later, a total

of 21 more specimens had been reported. Then, in 1933, came the extraordinary capture by I. T. Sanderson of 317 specimens of *R. sjostedti* in the Cameroons, and the following 20 years added about 40 more specimens, belonging to a dozen species.

Hence at one time it could be said that a probably unique feature of the Ricinulei was that they were an order well known for over a century, during which period not only every species but actually every specimen captured could readily be traced.

Such a statement of rarity can no longer be made. In 1966 my former pupil J. Pollock took about 150 specimens of *R. afzelii* in Sierra Leone, and in Guyana the late G. P. Lampel in 1959 collected over 50 specimens of two new species. More remarkable results, however, followed the exploration of the Mexican caves by Mitchell and others. The American genus Cryptocellus had long been less well represented than Ricinoides, but between 1967 and 1973 specimens of the former were being found in hundreds. One cave, La Cueva de la Florida, yielded 1,035 specimens of *C. pelaezi*. In the same years the number of known species was more than doubled and now approaches 30.

Two features of the Ricinulei combine to make them the most romantic order of the Arachnida, and might even support a claim to be placed among the most absorbing orders of the animal kingdom. These are their discovery in fossil form before a living specimen was found; and the reputation for extreme rarity, which they retained for more than a hundred years.

Add to this the unusual number of their anatomical specializations, and the conclusion is unassailable that the unique Ricinulei are the most fascinating, the most intriguing and the most challenging members of the invertebrate world.

28

The Order Pseudoscorpiones

[Pinces Geoffroy, 1762; Faux-scorpions Latreille, 1817; Chélifères Gervais, 1844; Chernetidae Menge, 1855; Chernetes Simon, 1879; Pseudoscopiones Pavesi, 1880; Chelonethi Thorell, 1885; Chelonethida Cambridge, 1892; Pseudoscorpionides Beier, 1932; Pseudoscorpionida Petrunkevitch, 1955]

Arachnida in which the prosoma is covered by an undivided carapace, bearing not more than two pairs of lateral eyes. The opisthosoma consists of 12 distinguishable somites, the last reduced to a circumanal ring. The chelicerae are of two segments and are chelate; they bear on the distal segment a spinneret from which proceeds silk secreted by prosomatic glands. The pedipalpi are very large, are of six segments and are chelate; venom glands within open near the tip of the metatarsus or tarsus or both. The legs are of seven segments, the first two pairs distinguishable from the last two. There are terminal claws with an arolium between them. There is no pedicel and no telson. The manducatory apparatus is complex: the sternum is usually absent. Two pairs of spiracles are found on the third and fourth opisthosomatic somites.

The prosoma of Pseudoscorpiones is covered by a carapace, quadrate or triangular in shape and almost certainly formed by a fusion of the original sclerites. It bears the eyes, two or four in number, when present; but some false scorpions are blind. Sometimes there are no transverse markings or furrows, but often these are present and allow the carapace to be divided into four regions. The first of these is the portion anterior to the eyes and known as the cucullus. In some families the cucullus is not distinctly separate, but in the Garypidae and Cheiridiidae it is narrow and elongated. It is morphologically the same part as the usually perpendicular clypeus of spiders, and is probably also homologous with the distinct jointed cucullus of the Ricinulei. The posterior, thoracic, portion of the prosoma consists of three parts, called by Chamberlin

(1931) the ocular disc, median disc and posterior disc. The furrows which separate these regions may be vestiges of the primitive segmentation, but this is not certain. They do not appear to be related to the insertion of the muscles within (Fig. 90).

The primitive number of eyes is four. Many species have two eyes and many are blind, but these conditions appear to have been derived by losses from the original complement. The eyes themselves are always sessile, and are situated close to the fore-edge of the carapace, except when the cucullus is elongated, and the two on each side are always close to one another. Structurally the eyes are of the pre-bacillar type, as found in the indirect eyes of spiders. They possess a tapetum and shine by reflected light. No false scorpion ever possesses median post-bacillar eyes like the direct eyes of scorpions and spiders, a fact which is a definite characteristic of the order. In the two-eyed forms, the anterior pair of eyes are retained, the posterior pair lost. The anterior eyes, in general are directed forward in slightly divergent directions, the

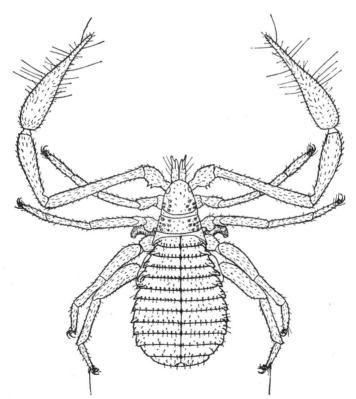

FIG. 90. Pseudoscorpion, dorsal aspect. Species, *Chelifer cancroides*.

posterior pair backward and somewhat upward. But false scorpions are not animals whose lives are much governed by the sense of sight; they are almost wholly dependent on touch.

In front of the carapace the chelicerae articulate. These are pre-oral appendages of two joints, specialized for four functions (Figs 91 and 92). Their most obvious use is comparable to that of the human hand—they hold the food which the animal is eating, and they pick up and carry food particles and also the grains of sand used in nest-making. They are spinning organs, bearing on the tip of the movable finger a galea or spinneret from which the silk issues. It is to this presence of a silk gland in the chelicerae that the false scorpions owe the name of Chelonethi, given them by Thorell. Thirdly, the chelicerae are bearers of sense organs in the form of setae and lyrifissures; and no doubt this variety of functions is the cause of the frequent cleaning of the chelicerae which the animals carry out when they preen.

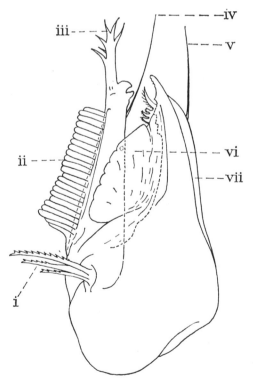

Fig. 91. Chelicera of a pseudoscorpion. After Chamberlin. (i) Flagellum; (ii) serrula exterior; (iii) galea; (iv) galeal seta; (v) laminal seta; (vi) lamina interior; (vii) lamina exterior.

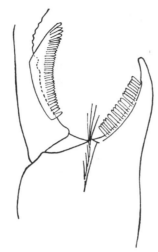

FIG. 92. Chelicera of *Obisium simile*. After Schenkel.

The chelicerae of Pseudoscorpiones are variable in size but constant in proportions throughout the order.

They may, as in the Feaellidae, be so small as scarcely to project beyond the edge of the cucullus, or they may be almost as long as the cephalothorax, as in the Chthoniidae. The first segment has a broad base or palm, prolonged into a pointed fixed finger. The second joint or movable finger articulates ventrally and moves up and down not quite vertically below the fixed finger. Upon the inner edge of each finger is a comb-like lamina or serrula. These are the active agents in the grooming of the pedipalpi, which are drawn through the "jaws" and so over the serrulae. When the chelicerae are large their tips generally cross when the movable finger is closed, the serrulae are attached to the fingers only for part of their lengths, and the galea is usually small or absent. This condition essentially fits the chelicerae to function as grasping organs. When the chelicerae are small, they are not so efficient for grasping. The finger-tips do not cross, but their tips fit into each other, the serrulae are attached throughout their whole length and the galea is usually well-developed. The earlier methods of classifying the order of Chelonethi were based on these differences. The spinneret or galea is always situated on the outer side of the tip of the movable finger. It is of two types, a short chitinous tubercle or a slender, translucent and often branched tube.

The silk glands with which it is in communication are of special interest, because they occupy the same position as the venom glands of many spiders. That two apparently similar glands should produce secretions so different in function and chemical composition might well suggest that they cannot be regarded as truly homologous, but their homology is made probable by the peculiar cheliceral glands of spiders of the genus Scytodes. These spiders capture their prey by a method which it is hard to parallel among other Arachnida. Approaching to within a few centimetres of their victims, they spit a mixture of silk and venom over them, a mixture that comes from the composite glands in the chelicerae.

Yet another important biological feature associated with the chelicerae is the flagellum, situated on the medio-ventral side, near the base of the inner digital condyle, and consisting in general of a group of specialized setae (Fig. 93). In more primitive families the number of setae is greater than in the more specialized, and varies from 12, in

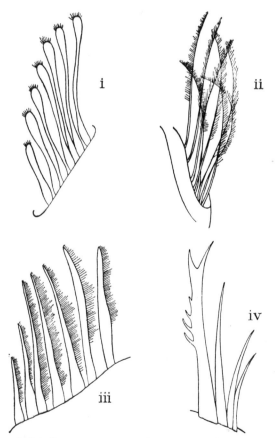

FIG. 93. Cheliceral flagella of Pseudoscorpiones. After Chamberlin. (i) *Hya heterodonta;* (ii) *Chthonius ischnocheles;* (iii) *Neobisium imperfectum;* (iv) *Atemnus oswaldi.*

some of the Chthoniidae, to one, in Geogarypinae. The individual setae are blade-like structures, feathery or toothed in appearance.

The interest of the flagellum lies partly in its function, which Vachon has described. When the fluids of the food have been sucked into the pharynx a solid residue remains on the labial setae; the chelicerae are drawn in and then thrust out, the setae of the flagellum spear the accumulation of debris and push it out between the coxae of the pedipalpi.

Further, a flagellum is found on the chelicerae of several other orders, providing a good instance of comparative arachnology.

In Solifugae the flagellum is limited to and is a mark of the male sex. It is generally but a single structure but it shows specific differences and is a valuable feature in classification (Figs 94 and 101). In the scorpions,

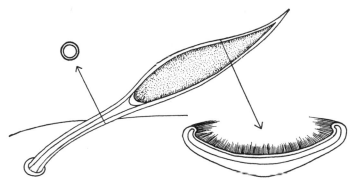

FIG. 94. Flagellum of male *Galeodes arabs*. After Roewer.

whose chelicerae are composed of three segments, the second segment is coated with setae on its inner surface. None of these is developed conspicuously into a flagellum, but their existence in this position is interesting, showing, perhaps, the primitive state from which the flagellum has been evolved.

In Araneae the proximal segment of the chelicerae is normally unarmed, but there are certain species in which it carries a spine or seta. The common British myrmecophile *Phrurolithus festivus* is the most familiar: a strong spine directed forward rises from the anterior surface of the paturon, quite close to the inner edge. Crosby (1934) has described a two-eyed spider, *Matta hambletoni*, in which the chelicerae are armed in front with a narrow short and blunt tooth (Fig. 95).

What appears to be the same thing is again to be found among the Acari, where the chelicerae of mites belonging to the genus Parasitus have a short spine at the base of the fixed finger (Fig. 96).

It must be admitted that there is at present no justification for the

FIG. 95. Left chelicera of *Matta hambletoni*, a Brazilian spider. After Crosby.

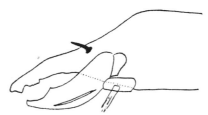

FIG. 96. Chelicera of the mite Parasitus, showing flagellum.

assumption that all these cheliceral appendages are homologous, although from their situation it may well appear that they are. Nor can they really be described as analogous, for their function is quite unknown.

In Solifugae the fact that the flagellum is confined to the male is suggestive of a sexual function, but this has not been observed on the rare occasions when the mating of Solifugae has been witnessed. Lamoral (1975) suggests that it acts as a store for an exocrine secretion, possibly functioning as a pheromone, which affects the female.

The pedipalpi are organs which are of the greatest importance, not only as the principal weapons but also as the bearers of many tactile setae. They also function in some species as secondary sexual organs. They are invariably six-jointed and chelate (Fig. 97). The first segment or coxa is provided with a gnathobase or manducatory process and is often called the maxilla. It bears a pair of delicate lamellae, essential components of the mouth parts. The femur and tibia are the longest joints and can be bent to a very small angle on each other. The metatarsus and tarsus compose the chela. The latter forms the movable finger and is articulated ventrally so as to work against the fixed finger above. It is opened by blood pressure and closed by a powerful adductor muscle which is spread within the swollen basal part of the metatarsus.

The form of the pedipalp is very diverse; so great indeed is the variation that it is scarcely an exaggeration to say that it is different for every

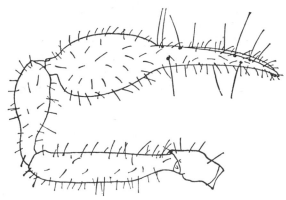

FIG. 97. Pedipalp of Chthonius. After Beier.

different species. There is generally a slight sexual dimorphism, the limb of the female being stouter than that of the male. The pedipalpi normally contain poison glands situated in the interior of the fingers with ducts opening at an orifice just below the tip of the last tooth. These glands may be present in both fingers or in either finger only or be absent altogether as, for example, from the Feaellidae. Just posterior to the poison tooth or venedens a blade-like modified seta, the lamina defensor, is found. Its function is unknown.

The metatarsus and tarsus together carry a number of tactile setae. These are long, simple and slender structures, each inserted at the bottom of a small depression. Ordinary "non-tactile" setae have not this depressed insertion. Typically each chela has 12 setae, though sometimes fewer and often more are present. The 12 are arranged in three series of four, one series on the exterior face of each finger and one on the inner face of the fixed finger.

The eight legs are naturally divided into four pairs, for throughout the order the forwardly-directed first and second pairs are different from the backwardly-directed third and fourth pairs. Owing to this distinction, Kästner (1931) has called the former Zugbeine and the latter Shubbeine, but as pseudoscorpions often run backward the names are not altogether appropriate. The first leg always closely resembles the second, but the latter is slightly larger and the third is similar to but smaller than the fourth. The legs of the posterior pairs are usually the longest and stoutest.

The legs are typically composed of seven segments, two of which compose the femur. There is no patella. An additional pretarsus, regarded by some as an eighth segment, is usually present. Modifications of this number of segments are found. Frequently the metatarsus and tarsus (telotarsus) fuse, forming a compound miotarsus, and sometimes the two parts of the femur are united. In the Heterosphyronida the forelegs have miotarsi and the hind legs have separate metatarsi and tarsi. In the Diplosphyronida all the metatarsi and tarsi are separate and in the Monosphyronida all have miotarsi.

The pretarsus consists of two claws and a membranous arolium between them. The latter is a short sucking pad, which enables the false scorpion to climb perpendicularly and to walk or rest on the underside of smooth horizontal surfaces. The tarsal claws are not toothed. In the forelegs of the males of most Cheliferidae the anterior claw is distorted and is slenderer than its fellow. The tip of the miotarsus is often produced into a distinct spine, which is used in mating to open the female genital operculum and assist the entry of the spermatophore.

The mouth parts of pseudoscorpions lie between the maxillae or gnathobases of the pedipalpi. The mouth itself is an aperture at the end of a tube or rostrum formed from dorsal and ventral projections. The latter, the labium or lophognath, is convex or crested dorsally, and is immovable. The upper lip, epipharynx or taphrognath, fits over the lower. By a slight upward movement of the taphrognath, the tube which they form can be enlarged, thus exerting a slight sucking effect. The true mouth, however, does not directly touch the food. The coxae of the pedipalpi enclose the rostrum in a trough-like chamber. Within this the laminae superiores and laminae inferiores lie closely below the rostrum, forming a compound tube which touches the prey in feeding and conducts the juices to the mouth. In the rostrum is an internal expansion forming a pharyngeal pump similar to, but not homologous with, the sucking stomach of spiders.

Weygoldt (1972) has described the action of the effective muscle, which in a Chelifer made 160 vibrations per minute. At intervals, peristalsis could be seen, forcing the food onwards from the mesenteron. Indigestible particles were ejected between the palpal coxae, which were then cleaned by rubbing them on the ground.

The ventral surface of the prosoma (Fig. 98) of pseudoscorpions is never protected by a large sternum, though a vestige may be found in certain species. This region is generally composed of the coxae of the legs and pedipalpi, which touch each other and meet in the middle line. However, in certain Chthonioidea a chitinous tubercle or platelet bearing one or two setae is found between the third and fourth coxae. This is a vestige of the true sternum. A secondary sternum is found in Garypidae, a mere membranization between the fourth coxae, and there is a larger secondary sternum in the small family Sternophoridae. The coxae extend further back than the carapace, and as a result the ventral surface of the first opisthosomatic segment is reduced.

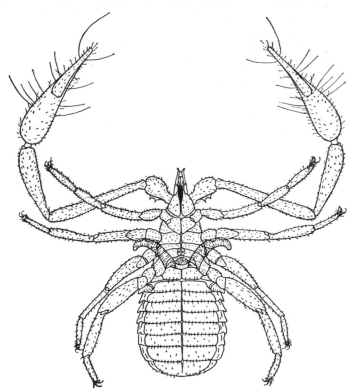

FIG. 98. Pseudoscorpion. Species *Chelifer cancroides*, ventral aspect.

The opisthosoma, more or less oval in form, is always broadly attached to the prosoma. It is fully segmented, with 12 tergites and ten sternites. The first sternite is either missing or fused with the second, which, with the third, forms the genital opercula. These enclose the genital chamber, which opens in the constant place on the second segment. The terminal somite is reduced to form the anal tubercle. Tracheae open on the third and fourth sternites.

These hard plates of the opisthosoma are united by an extensible membrane. Since the abdomen contains the digestive and reproductive organs, whose bulk is liable to change, the size of the opisthosoma varies a good deal according to season and circumstances. In many genera the tergites are medially divided into two parts.

The male organs within the second and third somites are an amazingly complex invagination of these two sternites. The slit leads to a uterus masculinus externus, and thence to a uterus masculinus internus.

From the former arise three large extensible chambers, the median and lateral genital sacs. The lateral ones are the protrusible "ram's horn organs" of Cheliferinae, which apparently function in sexual display. The median sac may serve as a seminal receptacle or as a reservoir for the secretions of the accessory glands.

The female genitalia are much simpler. The uterus receives the oviduct, and a pair of spermathecae also open into it. Legg (1975), in an analysis of the various forms of spermathecae, has pointed out that the possibility of storing spermatozoa "has enabled the pseudoscorpions to escape from the stable, static and humid habitat of soil and decaying vegetable debris and to exploit less favourable, non-static and temporary habitats". A median and a pair of lateral perforated plates, known as the cribiform plates, are found in association with the spermathecae. Their function is doubtful.

The bodies and limbs of all Pseudoscorpiones carry sensory spines and setae, and are well supplied with lyriform organs.

DISTRIBUTION

False scorpions are found everywhere except in the frigid regions of the Arctic and Antarctic. When the distribution of the separate families is considered some degree of localization is found, but several of them are cosmopolitan or nearly so. The following summary of family distribution is taken from Chamberlin's work.

FAMILY	DISTRIBUTION
Dithidae	West Indies, S.E. United States, Brazil
Chthoniidae	Cosmopolitan
Syarinidae	Rocky Mountains and Pacific Coasts of North America
Neobisiidae	Cosmopolitan
Hyidae	Philippine Islands
Ideoroncidae	Brazil, Paraguay, Mexico, W. United States, Sumatra, Siam
Olpiidae	Tropical
Menthidae	Deserts of W. Mexico and S.W. United States
Garypidae	Tropical
Feaellidae	Africa, India, Seychelles, Madagascar
Pseudogarypidae	Wyoming, Utah, Idaho, Oregon and California
Cheiridiidae	Cosmopolitan, except Australasia
Sternophoridae	Western Mexico and Australia
Pseudocheiridiidae	Nicobars, India, Burma and South Africa
Myrmochernetidae	Africa
Chernetidae	Cosmopolitan
Atemnidae	Oriental and Ethiopian, rare in America
Cheliferidae	Cosmopolitan

PALAEONTOLOGY

No fossil false scorpions of the earlier epochs have been discovered, but a fairly large number of species preserved in amber have been described. Some of these have been assigned to recent genera, and it is of particular interest to note that even in those remote times false scorpions had adopted the same method of securing dispersal that they use today. One specimen was found in amber attached to the leg of an ichneumon just as it might have been found at the present.

Amber-enclosed pseudoscorpions have come from the Baltic and from Burma, showing a wide dispersal comparable to the present condition.

A Miocene species, *Garypus birmiticus*, has also been described from Burma.

CLASSIFICATION

When Eugène Simon produced his account of the French false scorpions in 1879 he grouped them all in a single family, Cheliferidae, with three sub-families and six genera. Simple as this system was, it was the foundation of all modern methods.

In 1891 there first appeared, in a work by L. Balzan, the division of the order into two sub-orders, Panctenodactyli and Hemictenodactyli, based on the complete or partial attachment of the serrula to the chelicera. In this scheme there were four families and 13 genera, and it provided the foundation of all the systematic work on the group. It was greatly improved as a result of the labours of Hansen (1893) and With (1908) and in this modified form is adopted in the work of Berland (1932).

In 1931 there appeared the striking monograph of J. C. Chamberlin; the result of many years of intensive study, in which the whole order is subjected to a thorough scrutiny, and a full and comparative account is given of the structure of these Arachnida.

An alternative system, is that due to Beier in "Das Tierreich", 1933. He proposed a separation into three sub-orders, including 14 families with about 160 genera.

1 (2)   Legs 1 and 2 with tarsus undivided, legs 3 and 4 with tarsus of two parts CHTHONIINEA

2 (1)   All legs have similar tarsi, i.e. of one or of two parts (3)

3 (4)   All legs with tarsus undivided; eyes never four in number CHELIFERINEA

4 (3)   All legs with tarsus in two parts NEOBISINEA. (With the exception of the Feaellidae which have four eyes)

The most recent system is that of Weygoldt (1969), in which there are two "groups", three sub-orders and six super-families. In outline it reads as follows:

Group Heterosphyronida
  Sub-order Heterosphyronida
    Super-family Chthonioidea
      Families Tridenchthoniidae, Chthoniidae
Group Homosphyronida
  Sub-order Diplosphyronida
    Super-family Neobisioidea
      Families Neobisiidae, Gymnobisiidae, Syarinidea, Hyidae, Ideoroncidae
    Super-family Garypoidea
      Families Menthidae, Olpiidae, Garypidae
    Super-family Feaelloidea
      Families Pseudogarypidae, Synsphyronidea, Fenellidae
  Sub-order Monosphyronida
    Super-family Cheiridioidea
      Families Pseudocheiridiidae, Cheiridiidae, Sternophoridae
    Super-family Cheliferoidea
      Families Myrmochernetidae, Chernetidae, Atemnidae, Cheliferidae

The structural details on which the main parts of this classification is based are contained in the following key:

1 (2) The anterior tarsi of one segment, the posterior tarsi
      of two segments   HETEROSPHYRONIDA
2 (1) All tarsi with equal numbers of segments, either one or
      two   3
3 (4) All tarsi with two segments; four eyes   DIPLOSPHYRONIDA
4 (3) All tarsi with one segment: two eyes or none   MONOSPHYRONIDA

Few small animals have the same inherent fascination as have the false scorpions, and few naturalists, meeting them for the first time, fail to recognize their attractiveness. Their deliberate, almost pompous progress, alternating with rapid backward dartings, is like nothing else in the animal kingdom.

To the zoologist they present a combination of primitive features, like a fully-segmented abdomen, with specializations of their own, like cheliceral silk glands and complex methods of maternal care; as well as a cocoon-building habit which for instinctive capacity equals that of the spider in spinning its web.

All this leads to a rather surprising conclusion. We are sometimes told that, except statistically, there is no such thing as an average man, and this may be true, but in false scorpions we see an excellent example of a typical, average, arachnid. A wide distribution, a cryptozoic life, intricate structure, characteristic habits, mysteries and puzzles in plenty, all this builds up to the biology of a group of wholly delightful creatures.

29

The Order Solifugae

[Solpugides Leach, 1815; Solifugae Sundevall, 1833; Solpugae Koch, 1842; Solpugidea Cambridge, 1872; Solpugida Pearse, 1936; Mycetophora]
(Sun-spiders, Camel-spiders, Wind-scorpions; Walzenspinnen)

Arachnida in which the prosoma is covered by a propeltidium, marking a putative head, followed by three tergites. There are two, four or six eyes. There is no pedicel. The opisthosoma is of 11 somites, without a telson. The sternum is absent or hidden. The chelicerae are of two segments, chelate, powerful and projecting forward, with a flagellum in the male. The pedipalpi are of six segments, leg-like, and carry suckers on their tarsi. The legs are of seven segments, with the trochanters of the last three pairs sub-segmented, and the fourth pair carrying three to five malleoli or racket-organs on their coxae and trochanters.

The prosoma of Solifugae is as fully segmented as is that of any other arachnid. The anterior part consists of a conspicuously swollen "head", the large size of which is due to the muscles of the powerful chelicerae within it, covered by a single sclerite, the propeltidium (Fig. 99). Three small plates of chitin lie on each side below the edges of this; they have been called the lobus exterior, the lamina exterior major and the lamina exterior minor, and are particularly clear in the genus Rhagodes (Fig. 100). Behind the propeltidium lies the median plagula, followed by two paired transverse strips, the arcus anterior and the arcus posterior These show considerable variations among the different genera, and as a result different authorities have regarded them as having different relations to the underlying six somites of the prosoma. Finally, behind the arcus posterior there are two transverse quadrate tergites, the mesopeltidium and the metapeltidium. The divergent opinions mentioned are compared as follows:

ROEWER	KAESTNER	VACHON
1. Propeltidium	Propeltidium	Propeltidium
2. Propeltidium	Propeltidium	Propeltidium
3. Arcus anterior	Propeltidium	Propeltidium
4. Arcus posterior	Arcus ant. and post.	Propeltidium
5. Mesopeltidium	Mesopeltidium	Arcus ant. and post.
6. Metapeltidium	Metapeltidium	Mesopeltidium

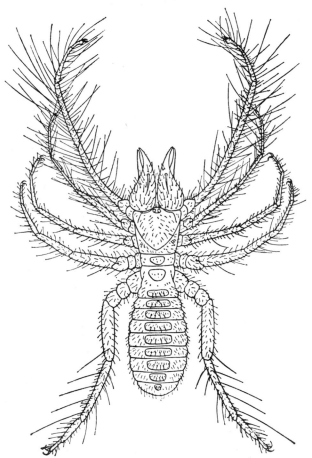

FIG. 99. Solifugae; dorsal aspect. Species *Galeodes arabs*.

Two eyes are placed on a small eminence near the fore-edge of the propeltidium.

It is evident that in Vachon's opinion the metapeltidium is to be regarded as the tergite of the first opisthosomatic or pre-genital somite.

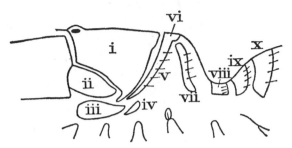

Fig. 100. Prosoma of Rhagodes: lateral aspect. After Roewer. (i) Propeltidium; (ii) lobus exterior; (iii) lamina exterior major; (iv) lamina exterior minor; (v) arcus anterior; (vi) plagula mediana; (vii) arcus posterior; (viii–x) tergites.

The opisthosoma has no pedicel, and consists of 11 clearly defined somites. Their tergites are undivided transverse plates, except the last, which is united to the last sternite to form a circumanal ring, a feature that is found in several other orders. The sternite of the pre-genital somite is a small triangular plate between the fourth coxae; some of the anterior sternites are partially divided by a split running forward from the hind-edge: the rest are undivided. The genital orifice is placed on the second sternite. It is a longitudinal slit-like aperture, peculiar among Arachnida in being guarded by movable lips, so that it can open and close. In several families (Eremobatidae, Karschiidae and Gylippidae) the female orifice lies transversely. The third, fourth and fifth show just behind their posterior edges the orifices of the opisthosomatic tracheae. There is a pair of these behind the third and the fourth sternites and a median one behind the fifth. The pleural membrane between the opisthosomatic sclerites is remarkably elastic and after a large meal the body of a solifuge may be swollen surprisingly. In this respect, as also in its general form, the body of Solifugae resembles that of the Pseudo-scorpiones. During pregnancy the increase in the size of the opisthosoma is immense; the stretched membrane seems to be almost at bursting point and the heavy mass to be almost beyond the animal's control.

There is great flexibility between the two parts of the body. During courtship the opisthosoma of the female is raised by the male; later the female appears to raise the prosoma. Flexion generally occurs posthumously, so that Solifugae preserved in spirit usually lie in this flexed position. This peculiar freedom is emphasized because it is a possible factor in determining the positions of the prosomatic sclerites. As the opisthosoma is raised there is a tendency for the last two prosomal tergites to be pushed forward, so that those directly behind the pro-peltidium get pushed to the sides.

The chelicerae of Solifugae are large and powerful, so large that they

may be as long as the rest of the prosoma, and so powerful that it is
probably true to say that Solifugae have the most formidable pair of
jaws in the animal world (Fig. 101). The chelicerae are of two segments
and closely resemble the very much smaller chelicerae of the Pseudoscorpiones. The first segment has a broad base or palm, continued into a
fixed finger. The second segment or movable finger articulates below
the base of this prolongation, and works against it in a more or less
vertical plane. The broad parts of the first segment are oval in section, so

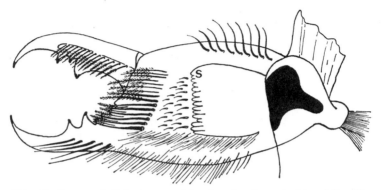

Fig. 101. Chelicera of *Galeodes aranoides*, female, from inside. After Roewer.
S, stridulating organ.

that there is a trough or channel-like space above and below them. The
former admits light to the ocular tubercle while the latter contains the
rostrum. The interior of the base contains the powerful muscles which
close the jaws and in well-fed specimens much fatty tissue as well. There
are no internal poison glands or ducts: the teeth are solid chitin and the
bite of Solifugae is not poisonous, but fatal only because of its severity.
Both fingers are provided with teeth, the number and position of which
is not only different in different genera (Fig. 102) but also differs in the
two sexes of the same species. In the mature males the teeth are often
reduced to a mere ridge, and it is doubtful whether chelicerae so modified can act efficiently for their true purpose. Many males die shortly
after copulation, but those of some species have been known to feed
and survive for at least 30 days.

A copious provision of spines and setae is also found on the surfaces
of the chelicerae. Between the two there is an important relation. When
they are in use, biting a struggling captive, the basal segments move
relative to each other in a longitudinal direction, while the mandibles
which work vertically, open and shut in alternation. This quickly

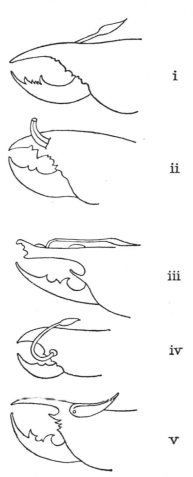

FIG. 102. Chelicerae of male Solpugidae, showing diverse forms of flagella. After Kraepelin. (i) Galeodes; (ii) Rhagodes; (iii) Solpuga; (iv) Hexisopus; (v) Daesia.

reduces the prey to a fluid state and the hardest parts are rejected unless the particles are very small.

They also possess stridulating organs. A smooth quadrangular surface is found on the inner side of each basal segment, where the two segments almost touch each other, and associated with this patch are spines of various stoutness. When the chelicerae are rubbed together a kind of twittering sound is produced, similar to that produced by stridulating spiders.

It is possible that these ridges serve two purposes, helping also in the grinding of the food; and this may account for the peculiar rotary masticating action of the chelicerae while the animal is feeding.

The chelicerae of male Solifugae also carry a flagellum (Fig. 103) in which respect the order differs from pseudoscorpions and others, where a flagellum is found in both sexes. There is no actual confirmation of the probable deduction that the flagellum has a sexual function, either in courtship or in copulation; and it has been said that if it is removed the animal is unaffected.

The rostrum of Solifugae is a very characteristic pointed beak, projecting from the ventral surface between the coxae of the pedipalpi. It consists of a hard lower portion and softer labrum above, the mouth being the space between them, and always carries a pair of sensory setae.

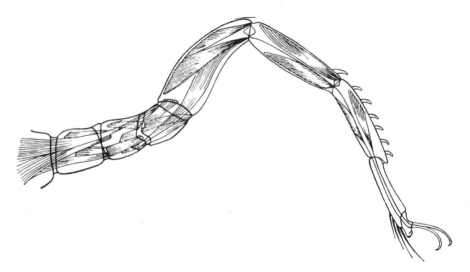

Fig. 103. Leg of Solpugidae, showing muscles. After Roewer.

The pedipalpi have the normal six segments. The coxae are provided with large gnathobases, which triturate the food just below the rostrum. On the ventral surface of each is a small tubercle bearing the opening of the excretory coxal gland. The trochanter is a single short ring-like segment, and the other segments are of normal form. The tarsus is very short. It ends not in a claw, but in a specialization peculiar to Solifugae, an adhesive organ. All segments of the pedipalpi are thickly covered with setae and spines of different lengths and strengths. Sometimes the arrangement of the spines is of use in classification, but sometimes it differs in the two sexes of the same species.

The pedipalpi are most active organs. They are used in picking up prey (often termites) both large and small, in fighting, including "stag-fights", in mating and in climbing smooth surfaces. In drinking, some species bring the tips together, thrust them into the water and raise them, bringing the liquid to the mouth like a pair of hands. Others, however, insert their chelicerae into the water and drink by suction. In battle the pedipalpi are invaluable; they determine the weak spots in the adversary's defence, and act as buffers to his attack. Provided as they are with long silky setae they act as very delicate organs of touch, and when the animal is running they are held out in front like the antennae of insects. The dependence of Solifugae on its pedipalpi is shown by the fact that if they are removed the legs of the first pair make clumsy attempts to act as substitutes in feeding.

The legs of the Solifugae are very characteristic. The first pair are the feeblest of all the appendages, but they are longer than the other legs and are not used as ambulatory limbs. They are carried stretched out in front and used as additional tactile organs, a habit found also in Amblypygi. This leg consists of seven segments, and its femur is divided into two; it ends in a tarsus of one segment. The second leg is a true ambulatory limb and ends in a tarsus of one to four segments. It has a divided femur, and the tibia and metarsus bear single or double rows of dorsal or ventral spines, much used in classification. In some genera the tibia has very characteristic projections at its dorsal end.

The third leg has a divided trochanter as well as a divided femur (Fig. 103). In general the spines and setae of the third leg resemble those of the second, but the spinal armature of the metatarsus is weaker than that of the second leg. The aberrant family Hexisopodidae is exceptional in this respect, as in many others. Between the second and third coxae is a broad area of softer epidermis, in which lies the opening of the prosomatic tracheae.

The fourth leg is the longest and strongest of all. It resembles the third in general, but the tarsus is often composed of more segments, and has a series of ventral spines. The coxae of this pair are inclined towards each other, while those of the second and third lie in the same transverse line; consequently, behind the fourth coxae there is a triangular space in which the genital aperture lies. Each leg of this pair carries five characteristic appendages, known as malleoli or racket-organs (Fig. 104). There are from three to five of these, on the coxae and trochanters, and their function is uncertain. Some zoologists believe that they have no function beyond providing support for the heavy opisthosoma, others that they receive vibratory disturbances of the ground. Since they are innervated, it seems certain that they are a type of sense organ, peculiar to the order. Whatever the truth may be, nothing like them is to be found in any other animal.

In the family Hexisopodidae, the form and arrangement of the legs

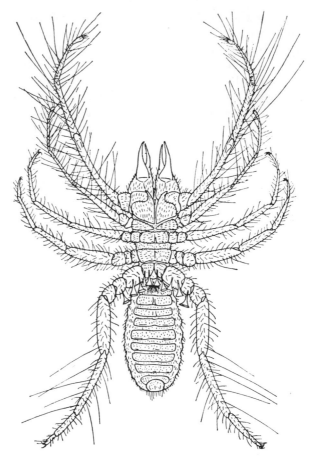

Fig. 104. Solifugae; ventral aspect. Species, *Galeodes arabs*.

differ from that of all others. The first pair of legs are much reduced and bear only three racket-organs each. Since the young of other families possess three such organs, this suggests that the Hexisopodidae exhibit neoteny. They and the family Rhagodidae have legs that are better adapted to digging than the legs in other families. The legs in all the Solifugae bear spines or setae of every degree of stoutness from short sharp spikes to long hair-like setae. There is not much evidence of thinking that any of these have any other function than that of organs of touch, as in all the orders of the Arachnida, but in Solifugae they seem to be more plentiful and more sensitive than in their allies. No other arachnid is so "hairy" that to a casual glance it looks like a ball of fluff, and no other can show such proportions as those found, for example, in the fourth leg of the male *Solpuga monteiroi*, where a leg 2 mm in diameter carries setae 3 cm long. It is said that if the tip of but a single seta be touched even with a hair, the animal responds instantaneously.

The opisthosoma of Solifugae is always a regular oval, the posterior margin being a smooth curve with no trace of a post-anal structure. It is always conspicuously and completely segmented, and the somites carry both tergites and sternites. The number of somites appears to be ten, but 11 are present in the embryo, and this number is really retained in the adult, the first somite being much reduced. Its sternite is a small triangular plate between the fourth coxae, and its tergite is similar in shape and lies hidden between the prosoma and the opisthosoma. The genital orifice on the second sternite is a slit-like aperture peculiar among Arachnida in being guarded by movable lips, so that it can open and close. The third, fourth and fifth sternites show just behind their posterior edges the orifices of the opisthosomatic tracheae. There is a pair of these behind the third and the fourth sternites and a single, median one behind the fifth.

DISTRIBUTION

Solifugae are confined mainly to the tropical and sub-tropical regions, and in Europe occur only in the south-east of Spain (Fig. 105). The distribution of the separate families may be summarized thus:

FAMILY	DISTRIBUTION
Melanoblossiidae	South-East Asia; South Africa
Eremobatidae	North and Central America
Karschiidae	Asia; Near East; north-west Africa
Rhagodidae	north-east Africa; Near East; south-west Africa
Hexisopodidae	South Africa
Gylippidae	East Asia
Solpugidae	Africa
Ammotrechidae	South and Central America; southern North America
Galeodidae	Asia; North Africa
Ceromidae	South Africa
Amacataidae	western South America
Daesiidae	Near East; Africa; Mediterranean zone

The above table may be expanded thus. Solifugae occur over almost the whole of Africa. In Asia they are to be found in Arabia, Persia, Turkestan and India. They also occur in Celebes and Indo-China. The Near East species extend as far west as the coast of Palestine and as far north as the Caspian Sea.

The American Solifugae fully studied by M. H. Muma belong to the families Eremobatidae and Ammotrechidae. They are now known from

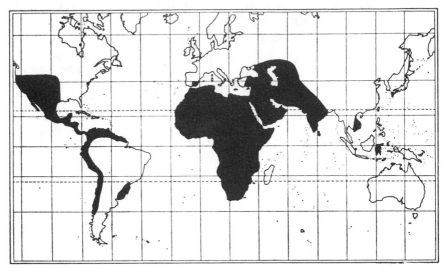

FIG. 105. Map showing distribution of Solifugae.

Terra del Fuego, throughout Argentina, from Uruguay and Brazil and all over the West Indies. In the north, records have come from Oregon, Washington, Idaho, North and South Dakota and Florida, Nebraska and Montana. In Canada Solifugae are known from southern British Columbia, Alberta and Saskatchewan.

It is clear from this that the facile description of the Solifugae as "desert arachnids" is an over-simplification. Most of the African species belong to the families Galeodidae and Solpugidae, and while the former are characteristic of the deserts north of the equator, the southern areas include many diurnal forms, as well as those of the forests and grassy districts.

PALAEONTOLOGY

One specimen only has so far been described as belonging to this order. This is *Protosolpuga carbonaria* (Fig. 106), which comes, in a rather poor state of preservation, from Mazon Creek, Illinois. It is a species 2.4 cm long, with a typical broad head and the large chelicerae characteristic of its order. The pedipalpi are strong and heavier than the legs, the second pair of which are unusually slender.

CLASSIFICATION

The first classifications of Solifugae were due to C. L. Koch and E. Simon. For many years the generally accepted system was that of K.

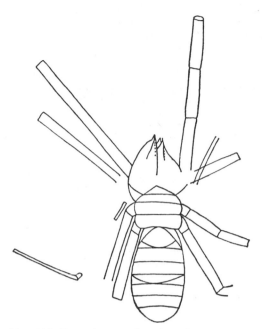

FIG. 106. *Protosolpuga carbonaria*. After Petrunkevitch.

Kraepelin in "Das Tierreich" of 1901. In this there were but three families:

1. Galeodidae
2. Solpugidae
   (i) Rhagodinae
   (ii) Solpuginae
   (iii) Daesiinae
   (iv) Eremobatinae
3. Hexisopodidae

But in 1901 only 164 species were known, whereas in the work of Roewer (1934) 595 species were distinguished and were arranged in ten families. Thirty years later the total exceeded 800, and two more families had been added. The following is a key to their separation.

CLASSIFICATION OF THE ORDER SOLIFUGAE

1	(2)	Male flagellum not paraxially movable	3
2	(1)	Male flagellum paraxially movable	17
3	(4)	Male flagellum of one or more setae	5

4	(3)	Male flagellum whip-like or membraneous	11
5	(6)	First, second and third tarsi of one segment; fourth tarsus of one to three segments	7
6	(5)	All tarsi of one segment; first tarsus with two claws	9
7	(8)	Female operculum not differentiated from other sternites	MELANOBLOSSIIDAE
8	(7)	Female operculum distinct from other sternites	EREMOBATIDAE
9	(10)	Leg 1 with two claws but no pretarsus; tarsi 2 to 4 of one segment; chelicerae with many small teeth; anus terminal	KARSCHIIDAE
10	(9)	Leg 1 with pretarsus and two claws; all tarsi of one segment; chelicerae with few teeth; anus ventral	RHAGODIDAE
11	(12)	Legs short and thick, fossorial; prosoma without visible tergites; cheliceral dentition greatly reduced	HEXISOPODIDAE
12	(11)	Legs long and slender, not fossorial	13
13	(14)	First tarsus with two claws	GYLIPPIDAE
14	(13)	First tarsus without claws	15
15	(16)	Second and third tarsi of one or two segments; fourth tarsus of six or seven segments	SOLPUGIDAE
16	(15)	Fourth tarsus with one to four segments	AMMOTRECHIDAE
17	(18)	Tarsal claws with micro-setae	GALEODIDAE
18	(17)	Tarsal claws smooth	19
19	(20)	Leg 1 with pretarsus and two claws; second to fourth tarsi of two segments	CEROMIDAE
20	(19)	First tarsus without or with only one vestigial claw	21
21	(22)	Male flagellum in two parts	AMACATAIDAE
22	(21)	Male flagellum in one part, membraneous	DAESIIDAE

IV. DE ARACHNOLOGIA

30

Economic Arachnology

The Arachnida, mites obviously excepted, are not animals that have attracted attention because of their evident economic importance. Indeed, it is possible to argue that the relative neglect from which they have suffered is due to this fact. Spiders in particular, the commonest and most obvious of the Arachnida, seem to do little direct damage beyond that of nuisance and are of little apparent value. Of the rest, scorpions in certain countries are poisonous or are at least an irritation, but the others are inconspicuous. As far as can be seen, every one of the Solifugae, Palpigradi, Opiliones, Pseudoscorpiones and Ricinulei might be simultaneously obliterated without any immediate effect on prices in the world's markets.

This, however, takes a narrow view of economic importance, limiting that term only to such creatures as may be described as damaging pests or obvious beneficial forms. When spiders catch flies, as they do in vast numbers, and when the other Arachnida capture whatever their prey may be, they are, one and all, playing their parts in the universal drama of life. They operate within the balance of nature and to this extent each one is important. For we know, only too well, how surprising may be the consequences when either thoughtless acts of men or natural cataclysmic events disturb this equilibrium, and how the accidental removal of some natural enemy may be followed by the almost un-limited increase of a pest species until subsequent happenings limit its reproduction and the balance is restored.

Recent studies have demonstrated that spiders feed upon, and may be capable of regulating, household pests, field crop pests, orchard pests and forest pests. Riechert has stated "Spiders clearly serve as stabilizing agents of invertebrate populations both in natural habitats and in the monotypic areas associated with agricultural lands". Who is to say that the same may not be true for those scorpions and Solifugae that feed gluttonously on termites, and pseudoscorpions that feed on Collembola and Psocoptera? Even the long-legged harvestman, so commonly condemned as useless, is a cog in the wheel and gives things a push in some direction. We do not know enough to say precisely where or which way, but it is certain that the natural world would be somehow different if the harvestman were not present. Therefore, let it be admitted that there is probably an indirect economic importance for all Arachnida, each one of which bears its share in this process.

The following are examples of direct economic importance for Arachnida other than Acari:

(i) Limulus has been used in America as food for both pigs and poultry and in some areas it is believed that this food makes the hens lay. Females are preferred on account of the eggs they carry. Thai (Siamese) hunt the species *Tachypleus gigas* for the sake of its eggs and a female has been sold for the equivalent of 5p. The eggs are said to look like caviare and to taste like potato.

(ii) Scorpions of several genera e.g. Buthus, Centruroides, Centrurus and Tityus, are known to be distinctly poisonous if not dangerous to man. Lay people in certain countries fear, avoid, or become mentally sick even from the stings of non-poisonous species.

(iii) Spiders of several genera e.g. Chiracanthium, Latrodectus and Loxosceles, are definitely known to inflict poisonous bites on man. Arachnid venom is given a fuller treatment elsewhere in this book.

(iv) Finally, although it is not apparent or well known, spiders of certain genera e.g. Nephila and Latrodectus, are in a small way beneficial to man. The silk of the former has been woven into cloth for ornamental tapestries and that of the latter has been used as cross-hairs in precision instruments e.g. surveying instruments, gun-sights and periscopes.

In the final analysis, however, the Arachnida whose known influence on human affairs is sufficiently important and far-reaching to justify this chapter are the Acari. They include species that attack growing crops, stored crops and manufactured produce; species which are parasitic on man and his domesticated animals; species which are hosts or bearers of pathogenic organisms and transmit them to man and his domesticated plants and animals; and species that attack, parasitize or transport themselves upon species that are beneficial or injurious to man and his industries.

The number of mite species is very great and their influence is chiefly felt through the large numbers of individuals which spring into existence wherever and whenever conditions favourable to their multiplication obtain. However, certain families contain species that are most frequently of economic importance. The work of the economic acarologist, like that of his brother, the economic entomologist, has consisted largely of biological and ecological studies on these animals, to determine the point at which they may be most successfully attacked or used. Mites do not collaborate in this process with much enthusiasm, for as Shipley said of lice "When you want them to live, they die and when you want them to die, they live and multiply exceedingly".

The following is a brief account of a representative selection of acarid pests. Since the subject is large and complex, no attempt at encyclo-paedic completeness has been made in this limited space.

Among those mites that attack and injure the growing crops of man, those of the families Tetranychidae, Tenuipalpidae, Tarsonemidae, Eriophyidae and Acaridae are the most important. The Tetranychidae, also known as red-spiders or spider mites, are probably the most fre-quently encountered mite enemies of the home gardeners. Although small, they multiply so rapidly that they easily become a serious economic menace. *Tetranychus urticae* (and *T. cinnabarinus* in green-houses) in England and Europe attack cucumbers, tomatoes, hops, strawberries, carnations, peaches and many other vegetable, fruit and ornamental crops. Trouble begins with the rising spring temperatures which awaken the hibernating pests and continue throughout the summer as successive mite generations feed and lay their eggs upon the stems and leaves of the plants. In October, the females are fertilized, change from dull green to brilliant scarlet, become gregarious, live in communal webs and finally hibernate or lay over-wintering eggs, completing the cycle of seasonal life.

Two other important pests of fruit crops should also be mentioned. They are *Panonychus ulmi*, the apple red-spider which is found on apples, plums, pears and raspberries, and *Bryobia ribis* found on gooseberries. These are leaf-feeders that suck the sap from the under-side of the leaves. In general they pass the winter as eggs.

In America, and on other crops, these and other genera of spider mites often become numerous and seriously injurious. For instance *Tetranychus pacificus, Bryobia praetiosa, Panonychus citri, Eotetranychus sex-maculatus* and other species may damage and defoliate apple, plum, almond and other fruit trees on the Pacific coast. Whereas *Eutetrany-chus banksi, P. citri* and *E. sexmaculatus* are the species most important on Florida citrus trees, *B. praetiosa* also seriously damages grass and clover crops and in autumn often enters houses in large, nuisance numbers.

Flat mites of the Tenuipalpidae are very small, flattened, red or red, black and green mites that seriously injure citrus, avocado and other fruits in the tropical and sub-tropical regions of the world. Perhaps the most important, especially in the Americas, is *Brevipalpus californicus* which is known to attack papaya, guava, citrus and avocado. In Florida, this mite is known to transmit or cause the disease condition known as Florida leprosis on citrus trees.

Although the tiny Tarsonemidae are known to include plant-feeding, fungus-feeding and insect-feeding species, perhaps the most important are *Polyphagotarsonemus latus*, the broad mite and *Stenotarsonemus pallidus*, the cyclamen mite. The former as the generic name implies feeds on and injures many different kinds of plants. In the Americas it is known to damage citrus both in glasshouses and in groves. The latter, despite its common name, probably achieves greatest importance for the damage it produces on glasshouse and field-grown strawberries. It should also be mentioned that tarsonemids of the genus Iponemus feed on bark beetle eggs, live as commensals in the galleries and transport themselves (phoresy) from gallery to gallery on the adult beetles.

Among the Eriophyidae, the rust, bud and gall mites, the genus Eriophyes contains a very large number of species, all of which are plant parasites. These curious creatures are barely 0·1 mm long, and invisible to the naked eye. The tropicopolitan citrus rust mite, *Phyllocoptruta oleivora*, is perhaps the best known and most important of the russet-producing species. It produces a mild to severe russeting or shark-skinning of citrus fruits throughout the world, is often responsible for small, early-maturing fruit and has been implicated in the transmission of citrus greasy spot. Many other russet-producing species are known, among them the apple rust mite, *Aculus schlectendali*, and the pear blister mite, *Eriophyes pyri*. The latter is well known as a pest all over England, Europe and America for its attacks on pear, apple and rowan. Leaves of infested trees open in the spring with small yellow or dark green spots that soon turn brown; the leaves then blacken and die. This is caused by mites which have wintered under the bud scales and burrowed into the leaves where they lay their eggs. The hatching young then spread the trouble by burrowing throughout the leaves and into other leaves. This is continued for the entire summer but before the leaves fall the mite is safely hidden in a bud for wintering.

There are also many bud and gall mites. Among them, in western North America, is the citrus bud mite, *Aceria sheldoni*, which causes serious abortion and loss of buds on lemon trees. In Great Britain, the species which most frequently demands the attention of fruit growers is the blackcurrant gall mite, *Cecidophyopsis ribis*, which is now also widespread in Europe and Canada. It produces "big bud" in blackcurrants and also attacks white- and redcurrants, flowering-currants and gooseberries. The presence of the parasite causes the buds to swell to almost double their normal size and die in the early weeks of summer. Red- and whitecurrant buds swell only if the infestation is particularly severe but they also die. Gooseberry leaves are often attacked and are undersized and yellowish. This mite can only be found by microscopic examination of parasitized buds between July and March when the mites are living and depositing eggs in the bud tissues. On the death of the bud, the mites move to the leaves and on bright days when a slight wind is stirring attempt migration by clinging to the legs of passing insects or springing upwards to be blown on the wind as aerial plankton. During heavy migrations, many thousands of individuals leave a bud in a day and little bunches of a hundred mites together may be seen rolling out of the buds on to the leaves below.

Blackcurrant is also attacked by two other gall mites. *Phyllocoptes masseei* and *Oxypleurites neglectus*, but these, although disfiguring, are not responsible for great damage.

Reference should also be made here to the large number of species of the extensive genus Eriophyes which are responsible for the formation of the apparently innocuous galls on many plants, including trees, shrubs and herbs. Mite-galls can be distinguished from insect-galls by a natural aperture through which the mites can escape. Mite-galls are also generally covered with a felty mass of "hairs" produced by the abnormally stimulated epidermal cells of the plant. Mites in the middle of this dense mass are often difficult or nearly impossible to detect. Examples of common mite-galls are the red growths seen on sycamore leaves, the round red galls on alder, the "nail galls" on lime trees and the galls on lichens of the genus Ramalina. The "witches brooms" or bundles of small twigs on the branches of birch, hackberry and other trees are apparently the result of a symbiosis between mites and fungi. Interesting as this widespread mite-gall parasitism is, it cannot claim more than passing mention, since its economic influence is insignificant.

One would also be remiss not to mention here the acarid, *Rhizogly-phus echinopus*, commonly known as the bulb mite. This species either causes damage or compounds insect and fungus damage to narcissus, jonquil, tulip, lily and similar plant bulbs. Mites of the family Acaridae are normally thought of as feeders on dying or dead plant and animal tissues, stored crops and manufactured produce, but the bulb mite apparently may cause primary as well as secondary injury.

Mites of the super-family Acaroidea including the families Acaridae, Carpoglyphidae and Glycyphagidae are perhaps our most important house and factory mite pests of stored agricultural crops and manufactured products. The cheese mites are undoubtedly the first to demand attention. Several species of mites may be found feeding on cheeses in factory, warehouse and home; *Acarus siro, Tyrophagus longior* and *Tyro-phagus putrescentiae* are three. Two or more species may often be found feeding and living together. Besides cheeses, these mites are able to live on many other stored and manufactured products such as hams, bacon, stored grains, breakfast cereals, flour and dried milk. They also feed on dead insects and the media of insect cultures.

These mites are often viviparous, a fact which saves time in their development with the result that in favourable circumstances they can multiply with extreme rapidity and soon completely consume a piece of cheese. When this change from food plenty to scarcity occurs, the oldest and the youngest members of the colony usually die. The rest transform themselves into flattened, round or eliptical forms known as hypopi, especially adapted for environmental resistance and population distribution. Each hypopus has a hard brown protective covering, very different from the globose, soft body of the active, non-migratory mite. The hypopi await the arrival of a fly, beetle or other arthropod to which they then cling for transportation to other surroundings.

Tyrophagus longior may also be found in bran and flour, as well as in cheese and *T. putrescentiae* in copra. *Glycyphagus domesticus* often infests sugar, and related mites occur in straw and hay. *Acarus siro*, the flour mite is a common pest in grain, bran, oatmeal, flour and cheese, where it is responsible for much damage. It bores a hole in the epicarp of the corn grain and eats the whole of the contents. The life history may, under favourable conditions, be completed in less than a fortnight so that it can multiply very rapidly. Although 50 species of mites had been recognized by 1948 as occurring in stored products of various kinds, many are now known to be predatory species that prey on other mites. Since all were detected in damp mouldy stores, the infestations may have been the consequence of spoiled produce rather than the cause. Control of stored-product mites may be obtained by drying the atmosphere of the store rooms. As the evaporating power of the air increases, the resistance of the mites is seriously diminished, free circulation of dry air is fatal, even to their eggs. Since hypopi of these mites are distributed on the legs of flies, beetles and other arthropods, all measures which tend to diminish or exclude arthropod infestations from the store are to be recommended.

Although acaroid mites are best known economically for their infestation of stored products, a few have been implicated in human and animal health problems. A species of Tyrophagus is believed to be the causal agent of human intestinal and urogenital acariasis. *Glycyphagus domesticus* is an intermediate host of a rodent tapeworm and at least one of the causative agents of "grocers itch" of food handlers. *Rhizoglyphus parasiticus* apparently attacks coolies employed on Indian tea plantations causing an irritation known as "water itch". *Tyrophagus longior* has also been indicted as the cause of "itch" in those that handle flour and is known to produce conjunctivitis. The copra mite, *T. putrescentiae*, attacks the bodies but not the faces of coolies and others engaged in the copra industry, causing intense irritation. Species of

Thyreophagus are believed to be predators of insects and other mites, and are often contaminants of grains and cereals.

Mites and ticks that attack humans and domesticated animals are often related to these Acaroidea, but are found in such families as the Psoroptidae and Sarcoptidae. Other families important in human and animal health problems include the Demodicidae, Dermanyssidae, Halarachnidae, Ixodidae, Pyemotidae, Pyroglyphidae and Trombiculidae. These nine families contain human and animal parasites that may be distinguished or classified as relatively innocuous forms, irritating forms, debilitating forms, lethal forms and disease transmitting forms.

Much the commonest of those that are relatively innocuous to man is *Demodex folliculorum* which lives in the hair follicles and sebaceous glands of the forehead, nose and chin. They may be discovered by pressing out the contents of a black-head pustule and examining it in a drop of xylol under a microscope. The mite is elongated with a tapering annulated abdomen and four pairs of tiny, strongly-clawed legs. Females lay oval eggs from which legless first larvae emerge. Although it is possible that this mite may assist in spreading a tropical skin disease called Lichen spinulosus, and has been implicated with a scalp dermititis, infestations usually cause no noticeable symptoms in man. However, the same cannot be said for *Demodex canis*, a mite that causes demondectic mange in dogs and cats. This mange is irritating, debilitating and in severe cases can be lethal. There is some question as to whether demodectic mange in dogs and cats is caused or aggravated by a Staphylococcus which occurs with the mite but this question is academic, since it is known that control or eradication of the mites from the mange lesions results in elimination of the mange. Recent studies have also shown that this mite works its way into the body and through the nervous system to the brain, so that additional uncorrelated damage may occur. Another species, *Demodex equi*, is suspected of causing pruritis in horses with a gross loss of hair.

Among the Dermanyssidae, the familiar "red mite" or chicken mite, *Dermanyssue gallinae*, is perhaps the commonest pest of cage birds and poultry. This species is about 1 mm long and feeds on the blood of birds while they are roosting at night. After filling themselves, they return to their hiding places on the bird-cage or hen-coop where their eggs are laid and hatched. Although the mites do not transmit any known disease the loss of blood is debilitating to the birds, causes loss of condition and in large infestations can cause death. Another species, the tropical fowl mite, *Ornithonyssus bursa*, causes similar serious damage to pigeons and other birds in tropical and sub-tropical countries. Human beings living under, beside or near pigeon lofts are often attacked by this mite, which leaves distinct, painful-feeling welts. Certain Dermanyssidae are also proven transmitters of disease.

Some parasitic mites seem to be wholly innocuous and of no economic significance. This appears to be true of the Halarachnidae. For example the occurrence of Halarachne in the lungs and air passages of seals and of Pneumonyssus in the lungs of Old World monkeys is only of interest in showing how widely mites have sought for favourable environment.

The opposite is true of the Ixodidae, and to a lesser extent the Argasidae. These large, hard or leathery-bodied ticks are not only irritating, debilitating and sometimes lethal forms, but also have the somewhat dubious honour of being the most versatile vectors of disease-causing viruses, rickettsiae, bacteria, spirochaetes and protozoa. Ticks attack a great variety of host animals, with fish the only major vertebrate exception, under a great variety of conditions; and seem to vary from region to region not only in their ability to attack various hosts, but also in the ability to transmit disease organisms. For these reasons, tick specialists, research scientists and medical men admit that much work remains to be done in the disciplines of tick taxonomy, tick zoogeography, tick biology, tick physiology and disease organism taxonomy and etiology. It seems, therefore, prudent that this presentation be confined to the broad aspects of tick-caused maladies.

Irritation by ticks is, of course, more or less obvious. It varies from the simple discomfort of individuals either climbing and walking over the body or inserting their biting organs and sucking blood to the more complex irritations caused by the hundreds or thousands of ticks in large, debilitating infestations. Any or all of the ticks that attack man and domesticated animals are capable of irritation.

Debilitation by ticks involves several factors, any of which, under severe conditions, can result in the death of the attacked man or animal. The first factor is the direct removal of blood, exsanguination, which results in extreme host discomfort and lassitude and many cause anaemia and death of the host. Second is the effect of tick toxins, tick toxicosis, in hosts. Such may or may not be associated with and complicated by blood removal. Several ticks attacking man have been suspected or indicted in this respect; they are *Argas reflexus* in the Near East *Ixodes ricinus* in France and a species of *Hyalomma* in South Africa. In animals, *Ornithodorus lahorensis* has caused toxicosis of sheep in southern USSR, *Ixodes ricinus* a paralysis of foals in Azerbaijan, and *Hyalomma transiens* "sweating sickness" in cows and goats in South Africa. The third factor is tick paralysis, presumably caused by a neurotoxin. This condition is known from several parts of the world and may or may not be separable from tick toxicosis. Human tick paralyses have been reported involving *Dermacentor andersoni* in Canada, *Amblyom-*

ma americanum in the southern United States and *Otobius megnini*, a spinose ear tick, in South Africa. For animals, the incriminated ticks and areas are *Ixodes holocyclus* in eastern Australia, several species of Ixodes and Rhipicephalus in South Africa, *Dermacentor andersoni* in north-western United States and *Dermacentor variabilis, Amblyomma americanum* and *Rhipicephalus sanguineus* in eastern and southern United States.

As mentioned above, the most important and most complex tick injuries involve the transmission of disease organisms. In the space allotted here it is impossible to do more than mention and comment on a few of the better known diseases and their known or suspected tick vectors.

Among the tick-borne viruses, epidemics of Kyasanur Forest virus have affected large numbers of people and monkeys in south-western India. *Haemaphysalis spinigera* is considered the important vector. Colorado tick fever, transmitted by *Dermacentor andersoni*, is the most prevalent human tick-borne virus disease in the western United States.

Rickettsiae and perhaps rickettsial-like organisms responsible for spotted fever in the United States may be spread from rodents to human beings by several kinds of ticks, but *Dermacentor andersoni* and other species of that genus are at present regarded as the primary vectors. Q fever, a world-wide rickettsial disease, has now been isolated from a number of kinds of ticks, all of which could be vectors.

The bacterial etiological agent of Tularemia, *Francisella tularensis*, apparently has become adapted to several modes of transmission among wildlife, to domestic animals and to man. Several genera of ticks. Haemaphysalis, Dermacentor, Rhipicephalus and Ixodes, are capable of harbouring the pathogen and transmitting the disease. Both ixodid and argasid ticks of several genera have been shown to be vectors of the disease known as Anaplasmosis, but the microbiological relationships of the causal organism is still undecided. The coccus-like *Anaplasma* bodies apparently are either protozoa, rickettsiae or bacteria.

Perhaps the most important human spirochaete disease is African tick-fever or relapsing fever, known by African natives as Kimputu and by Madagascan natives as Paraponjy. The causal spirochaetes, apparently many varieties or species of Borrelia related to *B. duttoni*, remain virulent for many years in argasid tick vectors of the genus Ornithodorus including *O. moubata*. Spirachaetasis of livestock can also be transmitted by ixodid ticks of the genera Boophilus and Rhipicephalus.

Red-water fever or Texas fever is by far the best known and most important tick-borne protozoan disease of cattle, horses, sheep and dogs. Since several causal organisms of the genera Piroplasma and Babesia are involved, the disease is also known as piraplasmosis or babesiasis. As might be expected with such a widespread, multiple-host multiple-pathogen disease, several ticks of the genera Boophilus, Dermacentor and Ixodes are also known to be vectors. Related important animal diseases of the eastern hemisphere, of which the best known is east coast fever of cattle, are also caused by tick-borne, protozoan parasites of the genera Theileria, Gonderia and Cytauxzoon.

Psoroptid mites can be important economic pests of their mammalian hosts. Members of the genus Psoroptes are ectodermal parasites of cattle, goats, horses, sheep and several non-domesticated animals; the species that produces mange on sheep may produce death. Dogs and cats are often attacked by the ear mite, *Otodectes cynotis*, heavy infestations of which result in fever and depression. Chorioptes species produce chorioptic mange at the base of the tails and on the legs of cattle, goats, horses and sheep, intensely irritating the animals.

The pyemotid, *Pyemotes ventricosus*, not only causes "grain-itch" among granary and cereal crop workers but may also complicate the natural control of stored grain pests and has been reported living within corn stalks. "Grain-itch" is a severe, eruptive dermatitis of grain handlers that can be quite irritating. Also, the mites are known to be parasites of larval and pupal Lepidoptera, and so may aid in the control of certain stored products pests. At the same time, they are also known to feed on parasitic wasps and thus may reduce natural control of certain pests. Since the mite has also been recorded from corn stalks, where it was apparently feeding on the plant juices, the species could be parasitic on the crop, could increase or decrease natural control of other pests, and finally complicate harvesting.

Certain Pyroglyphidae, particularly those of the genus Dermatophagoides are commonly found in house dust. One species, *D. pteronyssinus*, has been implicated as causal agent of house dust allergy, and was indicted as the cause of an irritating nasal and scalp disorder.

Rather more serious are the sarcoptic "scab mites or itch mites" of the genus Sarcoptes. The well-known *Sarcoptes scabiei* is the cause of human scabies, a dermatitis recognized by scabby, itching lesions that may persist for some time. Females, minute grey, tortoise-like mites, about 330 to 450 μm long, produce the lesions by burrowing under the skin in soft moist areas between the fingers and toes, around the knees, elbows and wrists and around the pelvis and buttocks. Scabies is often highly contagious in areas of high population density. Several similar mites, other races or other species of the same genus, are responsible for scabies of domestic animals such as pigs, sheep, dogs and cats. Sarcoptic mange of pigs and dogs can be particularly severe, debilitating or lethal.

The so-called "harvest mite" or chigger is perhaps the best known of all the mites that are parasitic on man. Trombiculid mites of several genera cause the well-known welt-like dermatitis or "trombidiosus" of chigger attacks. In the whole of Europe, the species most frequently involved is *Trombicula autumnalis*: in North America, species of Eutrombicula are most frequently indicted. Adult Trombiculidae are predatory on other arthropods and are most commonly found on or in the soil; it is the six-legged larva that is parasitic on vertebrates, including man. Before the true relationship of the larvae and adults were known, the larvae were often described under separate names. The pink to bright red welts with a central feeding tube or "stylostome" at the point of the chigger attachment is characteristic of chigger attacks. However, the amount of irritation produced by chiggers varies; some people have little or no symptoms, whereas others may even run a fever. These mites and other genera of the family also attack other mammals, as well as birds, reptiles and amphibians.

Before concluding this discussion on the economic importance of mites it should be pointed out that the field of economic acarology is relatively new when compared with that of economic entomology. In fact, for many years the problem of controlling the limited known number of economically important mites was considered within the province of the economic entomologist. However, as mites have become better known and the number of species has become more adequately differentiated it has become evident that the study of acarology demanded a special field of investigation. To this end, there are today mite specialists, taxonomic mite specialists of certain families and certain genera, mite physiologists and even economic acarologists. Since this latter expertise is rapidly growing and expanding it is no longer possible to give simple control measures for economically important mites. The reader is, therefore, referred to special economic treatises and texts on mites for adequate up-to-date chemical control measures for the species mentioned above.

The same applies to the so-called "biological or natural" control measures. Less than 20 years ago, it was felt that few if any mites could be considered to be beneficial to man and man's industries. Today a number of species are known to be effective predators of pest mites and pest insects and there can be no question but that many other beneficial species will be discovered as the field of economic acarology grows to become an important recognized scientific discipline.

31

Historical Arachnology

The history of the Arachnida begins with the myths and legends of ancient times. Ovid preserved for us the story of Arachne's contest with Pallas Athene; the scorpion occupies a place in the Zodiac and has associations with Mithras of Persia and Isis of Egypt; both spider and scorpion played their parts in the rites of ancient China.

Aristotle (384–322 B.C.) omniscient philosopher and founder of modern scientific method, made in his writings nearly 50 references to different arachnids. Scorpions are described as producing living young, contrasting with the spiders, that lay eggs. "The scorpion" he wrote "is provided with claws, as also is the creature like a scorpion that is found among the pages of books"; the first known reference to the pseudo-scorpions. Spiders received much fuller treatment; he noted the apparent rarity of males, saw something of courtship and wrote freely about their webs and their habits.

Pliny (A.D. 23–79) was more eloquent about scorpions than about spiders. He mentioned the occasional appearance of freaks with two tails, and was the first to use the names Tetragnatha and Solpuga. In other works of the first few centuries of our era there are also to be read such names as Lycosa, Mygale, Rhax and Phalangium though it is not possible to be sure what animals these were.

Rome fell in A.D. 395, and for more than a thousand years the history of arachnology made little progress. Coelius Curio wrote in 1544 "Araneus, seu de Providentiae Dei" a sermon, and the first book to be concerned only with spiders. Pierre Belon (1517–64), whose travel-book was published in Paris in 1553, was the first since Aristotle to base his writings on his own observations. The well-known work of Thomas Moffett (1553–1604) was the first book in which *Araneus diadematus* was recognizably described and illustrated.

In the meanwhile, mankind's chief concern with spiders was confined to the notorious tarantula (see Chapter 34).

A period of about a century and a half, from 1650 to 1800, may be described as the time of adolescence of arachnology.

The great Robert Hooke (1635–1703), one of the pioneers of microscopy, recorded his observations in "Micrographia", published in 1665. One paragraph refers to "a crab-like insect" that he had noticed in a book that he was reading. His microscope showed him that it had "ten legs two of which were like crab's claws". His drawing makes it evident that he was looking at a specimen of *Cheiridium museorum*. He also described at some length "the Shepherd Spider" (*Phalangium opilio*), comparing it with a crab and calling it *Cancer aereus*, an air crab.

Hooke was followed by Martin Lister (1638–1712), compiler of the first list of British spiders and the author of the first book to be written about spiders in English. He was a physician who numbered Charles II and Queen Anne among his patients, and was elected FRS in 1672. He may be said to have founded the tradition, maintained in Britain almost until today, that our arachnologists be doctors, clergymen, chemists, schoolmaster, anything rather than professional zoologists.

His book, "Historiae Animalium Angliae", was published in 1678 and described 34 species of spiders and three harvestmen, nearly all of which are clearly recognizable. He was the first to notice the differences in spiders' eyes, and the first to describe their chelicerae and the swelling of the male palp. He made the first reference to the little red mites that cling to harvestmen's legs.

Contemporaries of Hooke and Lister were Jan Swammerdam (1637–80) and Anthony van Leeuwenhoek (1632–1723), in Holland. The former discovered the keen eyesight of jumping-spiders and the attachment of female wolf spiders to their cocoons, and, more important, he wrote the first description of the use of the palpal organ in male spiders. Leeuwenhoek looked through his microscope at the claws on spider's legs and at their chelicerae, and he gave us the first reasonable description of their spinnerets. The period of adolescence ended with the work of Clerck in Sweden.

Carl Alexander Clerck (1709–65) was a civil servant in Stockholm. In 1737 he attended a course of lectures given by Linnaeus, that so roused his interest that he made the acquaintance of the lecturer and devoted his leisure to spiders and butterflies. In 1757 he published "Svenska Spindlar", in which he described 70 species of spiders, one harvestman and two false scorpions. The names that he gave to these were in theory to be replaced by those used by Linnaeus in the famous tenth edition of the "Systema Naturae" a year later, but because they had been so generally popular throughout Europe they were officially recognized by petition to the Zoological Congress of 1953. Clerck's book was widely welcomed and generously praised; it was, in fact, the first classic of arachnology.

Carl von Linné, later Carolus Linnaeus, (1707–78) was not an arachnologist, but his arrangement of the animal kingdom marks the beginning of systematic arachnology. He assigned 12 genera to the Insecta Aptera, and among these were Acarus, Phalangium, Aranea and Scorpio. It was from this beginning that the word Arachnides entered zoology, its author being Lamarck.

Jean Baptiste de Monet, Chevalier de Lamarck (1744–1829) professor of invertebrate zoology at the Museum d'Histoire Naturelle in Paris, began the task of splitting Linnaeus' large class Insecta. In his "Système des Animaux sans Vertèbres" of 1801 he created "Classe Troisième— Les Arachnides". This, therefore, marks the birth of arachnology proper, and the entry of French zoologists to the foremost place in arachnology, a position they have ever since retained.

First to follow Lamarck were P. A. Latreille (1762–1833) and C. A. Walckenaer (1771–1852). In 1802 Walckenaer produced a two-volume work on the fauna of Paris, in which he described 131 spiders, all in the same genus, Aranea. Within this genus he arranged the species in several groups, with the comment that their differences did not seem to be great enough to call for the creation of new genera.

Latreille, who was the first professor to hold a chair of entomology, was essentially a systematist, and between 1802 and 1829 he produced several works dealing with the classification of insects. In these the classification of Arachnides was steadily developed, and after Latreille's death the work was continued by Walckenaer. His "Histoire Naturelle des Insectes Aptères" 1837–47 is historically important because of the large number of still valid genera that were first described in it.

In the meanwhile, spiders were receiving attention in Germany, whence came the great 16-volume "Die Arachniden" of C. W. Hahn (1786–1836) and C. L. Koch (1778–1857). The existence of such an immense work played a great part in establishing in Germany a lasting interest in Arachnida.

Orders other than the spiders began at about this time to attract serious and critical attention. Carl J. Sundevall (1801–75) encouraged this wider view in his "Conspectus Arachnidorum" of 1855; William E. Leach (1790–1858) initiated the study of false scorpions; Alfred Tulk in 1843 made the first dissections of a harvestman; F. E. Guérin-Méneville described the first of the Ricinulei in 1838; H. Straus-Durckheim in 1829 pointed to relationship between Arachnida and Limulus. In Britain the study of spiders had scarcely advanced beyond the point reached by Lister in 1678. Its revival was largely due to Meade and Blackwall.

Richard Henry Meade (1814–99) was a surgeon practising in Bradford. In 1853 he privately printed a list of 231 spiders known in Britain, and two years later published a monograph on 15 British species of harvestmen. He frequently sought the advice of Blackwall, and incidentally performed a service to science by introducing Pickard-Cambridge to Blackwall in 1851.

John Blackwall (1790–1881) on retiring from business in Manchester, went to live at Llanrwst in Wales in about 1854. He had earlier abandoned the study of birds in favour of spiders, on which his first paper, dealing with gossamer, appeared in 1827. Virtually alone, he continued to investigate the spiders of Britain, finding, as Lister had never done, that most of the British spiders are very small animals. He also recognized the value of male palpi as a means of distinguishing species. In 1851 he began the production of a list of 210 British species, from which he developed his great "History of the Spiders of Great Britain and Ireland", published by the Ray Society in 1861 and 1864. This book is one of the great classics of arachnology, and shows that 70 British species, or nearly one-eighth of the whole, were discovered by Blackwall.

His work suffered in only one respect. His isolation left him unaware that Menge in Danzig and Westring in Gottenburg were simultaneously writing accounts of German and Swedish spiders, and in consequence there was much overlapping and the making of avoidable synonyms.

Octavius Pickard-Cambridge, who was mentioned above, was rector of the Dorset village of Bloxworth from 1868. Always a keen naturalist, he was inspired by Blackwall to study spiders, and this he did with exceptional success. From 1880 to 1914 he produced annual papers on new and rare spiders, and continued the work of Meade on harvestmen and wrote a monograph on false scorpions. His great work, "The Spiders of Dorset", was published in 1879 and 1881, and remained the standard description of British spiders for over 60 years. Pickard-Cambridge was elected to the Royal Society in 1887. He was, indeed, more than a famous arachnologist; he had a wide knowledge of other branches of natural history and he was an influential member of the Church of England in Dorset. Above all, he was the well loved leader of his flock for more than half a century.

Before Pickard-Cambridge died he had expressed the hope that his successor in Britain would be A. R. Jackson.

Arthur Randall Jackson (1877–1944) was a doctor of medicine in Chester, whose attention was directed to spiders by Professor Herdman:

spiders quickly became the chief occupation of his leisure. He had many of the ideal qualifications for an arachnologist, a scientific training, a keen eye for important detail, a gift for identification and the energy of a successful collector. Undoubtedly his greatest service to the science was the ungrudging help that he gave to all who appealed to him for assistance. This generosity had its tragic side, for it absorbed the time that might otherwise have been given to the writing of the badly needed book on the spiders of Britain. Yet it found its reward later in the work of those to whom he had passed on his knowledge and technique: the writings of Bristowe, Locket, and Millidge may, in effect, be regarded both as his memorial and as the sources of the widespread interest in Arachnida in Britain today.

Among the contemporaries of Jackson were Warburton, Hull and Kew.

Cecil Warburton (1854–1958) spent the whole of his professional life as a lecturer in Cambridge University. He was one of the first to make a careful search for spiders in the famous collecting area of Wicken Fen, and later he investigated the silk glands of Araneus and described the life cycle of Agelena. In 1909 the Cambridge Natural History contained his contribution on Arachnida, chapters which for many years were the recognized authoritative source of information readily available to English readers. After the First World War he transferred his attention to the Acari, and at the Molteno Institute took an important share in the production of an exhaustive treatise on ticks. He retired from his post of zoologist to the Agricultural Society when he was 92, and at the time of his death at 104 was probably the oldest man in Britain.

The Rev. J. E. Hull (1863–1960) was a north-country clergyman with a special interest in spiders and mites. The minute spiders of the large family Linyphiidae came especially to his notice, and much of his work was concerned with their names and classification, as well as with the description of new species.

Harry Wallis Kew (1868–1930) was for many years the chief authority on the false scorpions of Britain and Ireland. He was the first to describe their courtship, and to show how their different instars could be distinguished. In this and other ways he was a pioneer in the acquisition of precise knowledge of an unusually interesting group.

The influence of Jackson, Hull, Kew and others lives on in the continuing and increasingly successful work of those who were inspired by their example and helped by their advice. The results are to be seen in the energy and enthusiasm that secured the foundation of the British Arachnological Society, and in the existence of an up-to-date three-volume work on British spiders, of such quality that the names of G. H. Locket, A. F. Millidge and P. Merrett are unquestionably assured of immortality.

Significant as were the efforts of Blackwall and Pickard-Cambridge in Britain, more important events were taking place in France.

Eugène Simon (1848–1924) realized from a comparison of Walckenaer's book with that of Hahn and Koch how great were the possibilities in a study of the Arachnida of France. Quickly he passed to the spiders of the world, a project truly sufficient for a man's lifetime.

In pursuit of his ambition Simon became a traveller and collector, accumulating specimens wherever he went. These and the innumerable spiders sent to him by correspondents all over the world he described in annual papers, while at the same time he was at work on two encyclopaedic publications, his "Arachnides de France" (1874–1937) and the "Histoire Naturelle des Araignées" (1892–1903). The industry and devotion that made such an achievement possible have seldom been equalled. Simon received universal recognition as the supreme authority on all matters connected with spiders. Unlike most zoologists, he had also to create a systematic classification as he worked, while he surprised everyone by his phenomenal memory of all the spiders that had passed through his hands. His publications numbered 319 titles and his collection was contained in over 20,000 numbered tubes.

A career of this kind was bound to influence others, who were ready to carry on the work of the master. Thus there existed in France after the Kaiser's War a nucleus of zoologists from whom arose a French school of arachnologists that has ever since included some of the leaders in our subject.

Louis Fage (1883–1964) was a marine zoologist at Banyuls who, moving in 1920 to Paris, added spiders to his already broad zoological interests. He specialized in the study of Arachnida in caves. Among his contemporaries was Lucien Berland (1888–1962), an entomologist who was attracted by the Arachnida. His concern was largely with the arachnids of islands, and he made valuable contributions to zoogeography. In addition he wrote a full, general account of the class in "Les Arachnides" in 1932, and this, with other shorter books, did much to extend the general interest in spiders and scorpions.

On these foundations developed the great advances in Arachnology that we owe to such living zoologists as Bonnet, Millot and Vachon. It is impossible to refrain from a mention of Bonnet's monumental "Bibliographia Araneorum" which surveys all publications about spiders from the earliest times until 1939. Equally there must be recorded the great organization Centre International de Documentation Arachnologique, which under the care of Max Vachon, brings together the arachnologists of every nation.

The early work in France of Latreille and Walckenaer was paralleled in Germany by the efforts of C. W. Hahn (1786–1836) and C. L. Koch (1778–1857), who together produced an immense work, "Die Arach-niden", in 16 volumes. Described by Bonnet as "une merveille pour l'époque", it started German arachnology in a manner which it has ever since maintained. Koch was followed by his son, L. C. C. Koch (1825–1908), who was responsible for "Die Arachniden Australiens", and later by A. Menge (1808–80) the author of "Preussischen Spinnen", by P. Bertkau (1849–94) and F. Dahl (1856–1929).

During the present century the arachnology of Germany has been marked by the brilliance of individuals, such as U. Gerhardt (1871–1950) A. Kaestner (1901–71) and A. M. F. Wiehle (1884–). It has further been notable for the production of encyclopaedic works on the largest scale. "Das Tierreich", in which the contributions of K. Kraepelin to classification were conspicuous, and the immense "Handbuch der Zoologie" of Kukenthal and Krumbach, in which all orders of Arach-nida received the fullest possible treatment from a number of authori-tative writers. On the same scale as these were the personal publications of C. F. Roewer (1881–1963), whose monumental "Die Weberknechte der Erde" described the harvestmen of the world, and whose "Katalog der Araneae" listed the known species of spiders.

Arachnology spread steadily across Europe with activities in each country which rose and fell and rose again according to the influence of individual zoologists. It is impossible to refer to all those who have given us accounts of the faunas of their own lands, and it is undiplo-matic to mention but a few. Nevertheless, with this admission, it is impossible to pass in complete silence over the names of E. Reimoser (1864–1940) of Vienna, L. Caporiacco (1901–51) of Italy, S. A. Spassky (1882–) of USSR, M. Beier of Vienna and G. Kolosvary of Budapest. Beier's lifelong work on false scorpions and Kolosvary's book, "Die Weberknechte Ungarns", on harvestmen have deserved all the admiration that they have received.

Also to be especially noticed was the contribution from Denmark of H. J. Hansen (1855–1931) and W. Sørensen (1848–1916) in their book "On Two Orders of Arachnida". This was a much needed review of the then known species of Ricinulei and Cyphophthalmi, and did much to rescue two interesting groups from possible neglect and to provide a basis for all future research.

From Sweden, too, came one of the greatest of Arachnologists, T. T. T. Thorell (1830–1901). Thorell was a zoologist of wide interests guided by the soundest scholarship, and was the author of over 50 papers and books on spiders. In the taxonomy of the order he found a fertile field for his work, and his fine books, "On European Spiders" of 1869 and "On the Synonyms of European Spiders" of 1870 helped greatly to correct and stabilize a somewhat chaotic nomenclature.

He also surveyed the whole class, and was responsible for the founding and naming of several of the orders that are recognized today.

After a few tentative beginnings in the seventeenth century the arachnids of America attracted some notice by reason of Mlle S. Merian's well known account in 1705 of a large spider that ate small birds, an observation that was at first discredited and later verified by H. W. Bates.

Some early accounts of species by J. C. Fabricius and J. T. Abbot were followed by "The Spiders of the United States" by N. M. Hentz (1797–1856). Hentz was a schoolteacher with a medical training, and he is usually regarded as the founder of serious arachnology in America. He was followed by Emerton, McCook and Peckham.

J. H. Emerton (1847–1931) studied Arachnida for more than 60 years and wrote voluminously about them. He edited Hentz's posthumous works, was one of the first to study the spiders of the Arctic, and, perhaps most noteworthy of all, he travelled to Europe and made the acquaintance of his contemporaries in France, Germany and England. His contributions to our science have always been held in high esteem.

The Rev. H. C. McCook (1837–1911) achieved immortality by the writing of a large three-volume book, "American Spiders and their Spinningwork", which he published himself between 1889 and 1894. At that time there was no comparable work in the English language.

C. W. Peckham (1841–1914) made special studies of the spiders of the family Salticidae, and introduced the world to the rich field of spider behaviour, including their courtship and mental powers. Much of his work was published in collaboration with his wife, one of the first of the rather small band of women arachnologists.

In more recent times the study of Arachnology in America has been encouraged by the efforts of N. Banks (1868–1953), J. H. Comstock (1849–1931), T. H. Montgomery (1873–1912) and J. C. Chamberlin (1898–1964).

J. H. Comstock was a professor of zoology at Cornell. He was the author of an admirably produced and well-illustrated volume "The Spider Book" of 1913. It was valued by all arachnologists of that date for it included the orders of scorpions, false scorpions and harvestmen, and was a welcome supplement to Warburton's writings. A second edition was brought up to date by W. J. Gertsch and published in 1948.

J. C. Chamberlin of Stanford University was one of the great specialists in the study of false scorpions. Attracted to them as a boy, when very little was known about them, he produced a logical system of the order, and in addition to many papers summarized his system in a great monograph "The Arachnid Order Chelonethida" in 1931.

For many years arachnologists the world over looked to Alexander Petrunkevitch (1875–1964) as the supreme authority on all questions of arachnid evolution and classification. Born in Russia, he travelled to America in 1903 and in 1910 joined the University of Yale. His attempts to compile a satisfactory classification of spiders, his investigation of extinct fossil species and his later work on spiders in amber were the chief items in a long life characterized by an incredible daily schedule of hours in the laboratory, remarkable technical skill and an astonishing command of about ten European languages. The influence of his contributions to arachnology will survive for many years to come.

Africa, with its great fauna of conspicuous scorpions and Solifugae, has always offered boundless opportunities to arachnologists. After a few casual references to spiders among various descriptions of insects there appeared in 1825 the well known "Description de l'Egypte" by J. S. Savigny and V. J. Audouin. Nine plates, engraved by Savigny, accompanied descriptions of 29 genera of spiders, together with some account of other orders. Indeed, the history of both scorpions and Solifugae makes constant references to the African species; and from Africa too came in 1838 the recognition of the first of the Ricinulei, the species which at that time was known as *Cryptostemma westermannii*.

Most of the study of the African Arachnida has come from the southern half of the continent. The pioneer work must be credited to W. P. Purcell (1866–1919), J. Hewitt (1880–) and E. Warren (1871–1935), all of whom were born in Britain and spent a large part of their lives in Africa. Employed in the service of different museums, they were the pioneers of arachnology in Africa.

Here it is impossible to omit mention of a living arachnologist, Dr R. F. Lawrence. In 1922 he accepted a post in the South African Museum at Cape Town, where Purcell's large collections were preserved. In the following years he undertook extensive expeditions to various parts of the continent, returning, whether from mountain, forest or desert, with a full harvest of specimens. In 1935 he succeeded Warren as Director of the Natal Museum at Pietermaritzburg, and in 1953 published his impressive volume on "The Cryptic Fauna of Forests". Lawrence's work has covered almost all the orders of arachnids, and he must be recognized as having securely laid the foundations of African arachnology.

In the Near, Middle and Far East, arachnology has made steady if unspectacular progress during the present century. In these regions, as indeed in most parts of the world, its growth has followed the same course. At first a few curious examples of one order or another have been rather vaguely mentioned in a book of travels, and from these there grew some interest in the local fauna. More serious collections were made, often by explorers or professional collectors, and were brought home to specialists in Britain, France or Germany. In this way a very large number of foreign Arachnida were described by zoologists such as R. I. Pocock (1863–1947), F. H. Graveley (1885–), W. R. Sherriffs (1881–) and H. R. Hogg (1851–1923).

This period overlaps the later stage of development, when the inhabitants of a country, provided with museums and universities of their own, undertook these duties themselves. By progress of this kind there has grown an increasing concern with Arachnida in India, Palestine, Malaysia, Japan and elsewhere.

In the Antipodes the study of the spiders of New Zealand began with a collection made in 1827 during the voyage of the Astolabe. This was followed during the next 50 years by a small number of other collections, one of which included the species locally known as the Katipo, notorious for its poisonous bite. This was a species of Latrodectus, a relation of the famous black widow.

The true founder of New Zealand arachnology was A. T. Urquhart (1839–1916), a keen naturalist who made good use of his travels in the country. His many papers were published between 1882 and 1897, and together they gave a sound start to the subject. Some years passed before another took up the study. This was Comte R. de Dalmas (1862–1930). At first he brought his many specimens to Simon for report, but later he examined and described them himself.

The thorough and systematic description of the Arachnida of New Zealand, now in the hands of R. R. and L. M. Forster, is of great significance, since it has shown how the taxonomy of Arachnida has been handicapped by the fact that all systems have been based on specimens from the northern hemisphere, to the exclusion of the often very different fauna of the other half of the world.

The first Arachnida from Australia were briefly mentioned by J. C. Fabricius in his "Entomologica Systematica" of 1792–96. Subsequent collections were described by L. Koch and Graf von Keyserling in a great work of nearly 1,500 pages, "Die Arachniden Australiens". In the present century many Australian species were examined by H. R. Hogg. The first resident in Australia to continue this work was W. J. Rainbow (1856–1919). Born in Yorkshire, he had lived in New Zealand and Australia from 1873, and 20 years later became entomologist to the

Australian Museum. He wrote over 70 papers dealing with different aspects of the arachnida of Australia, and in 1911 summarized the whole in a great "Census of Australian Araneidae".

Nothing comparable with this has appeared more recently, but K. C. McKeown's "Australian Spiders" of 1952, a revision of his "Spider Wonders of Australia" of 1936, should not be overlooked.

Meanwhile the spiders of Tasmania were revealing the possibilities that awaited the zoologists of the island. V. V. Hickman holds the distinction of being the first arachnologist of the Commonwealth to hold a university chair, and his discoveries among the minute spiders on Mount Wellington stand in a class by themselves.

An historical chapter must stop short of the present generation, but it is legitimate to add that the advance in arachnology during the past 30 years has been rapid and worldwide. An important degree of co-operation between arachnologists is now in being, and has encouraged and directed the formation of organized societies, some information of which is of value.

1. *Toa Kumo Grakki:* Arachnological Society of East Asia
    Founded 1936. President: Dr Takeo Yaginuma
Membership is confined to residents in east Asia; arachnologists of other regions can subscribe to the Society's journal, *Acta Arachnologica*, which is issued twice a year. Its Japanese contributions are followed by English summaries.

2. *Centre International de Documentation Arachnologique*
    Founded 1950. Secretary-General: Professor Max Vachon
The Centre covers the world, with appointed representatives in most countries. Its Annuaire contains the names and addresses of all arachnologists and its yearly Liste des Travaux Arachnologiques records recent and forthcoming papers and books. An International Congress is held normally every third year in a different country, an invaluable occasion that should never be missed.

3. *The British Arachnological Society*
    Founded 1969. President: Professor B. J. Marples;
        Hon. Secretary: J. R. Parker.
This Society grew through the British Spider Study Group from the Flatford Mill Spider Study Group, inaugurated in 1959 by G. H. Locket. In every respect an active and practical company, it arranges field courses several times a year in different areas. Its Annual General Meeting also changes its location, so that the whole country is covered. Its *Bulletin*, accompanied by a News Letter, is issued in March, July and September.

4. *The American Arachnological Society*
   Founded 1973. President: Dr Beatrice R. Vogel;
   Secretary/Treasurer: Dr Mel E. Thompson.

Ten years earlier a study group, The Arachnologists of the South-West, was founded under the leadership of Dr B. J. Kaston of San Diego; and from it grew the larger American Society. Its aims are the same as those of the other Societies mentioned, and its organization is elaborate. Its excellent publication, the *Journal of Arachnology*, appears in January, May and August.

32

Practical Arachnology

The practical side of arachnology falls into five parts: (i) the collection of living material in the field; (ii) the permanent preservation of specimens for the future; (iii) the anatomical study of their bodies in the laboratory; (iv) the observation under controlled conditions of the behaviour of the living animal; (v) the determination of the life history by raising individuals from the egg to the adult.

COLLECTION

This does not differ in any essential from the collecting of other small animals, and the most productive methods are the usual ones of beating, sweeping and sifting. All these elementary methods may be profitably adopted by the beginner taking his first steps in arachnology, or by anyone about to determine the general arachnid population of a region under investigation for the first time.

In a previous chapter an account of the ecology of Arachnida put forward a division of any such area into three zones: the ground layer, the shrub layer and the tree layer or canopy. An arachnologist who carries on his collecting in the spirit of an ecologist will recognize the advantages of this and adopt his methods accordingly.

Arachnida on the ground may be taken most readily by sifting. Handfuls of dead leaves, pine needles and other vegetable debris are gathered into a sieve made of fine-meshed wire netting and shaken over a sheet of newspaper or white macintosh. Numbers of cryptozoa descend. The more restful methods of hand-picking and stone-turning are not to be neglected.

Specimens revealed in these ways may be coaxed into a specimen tube, or picked up with a moist forefinger, or sucked into a pooter. This useful tool is easily made in a variety of designs.

There are also species to be sought underground, so that a little digging with a sheath knife may yield strange spiders and unexpected harvestmen. Grass roots thus lifted, with some of the earth, should be shaken in the sieve.

These methods may well be supplemented by the use of the Berlese or Tullgren-type funnel. This consists essentially of a vertical cylinder, inside which is a horizontal sieve. An electric light bulb is fitted into the lid of the cylinder and a funnel is placed below it. The funnel leads into a bottle of alcohol. Masses of vegetation are put into the cylinder and the light is switched on: the animals tend to move away from the light and the dryness of the topmost region, until they pass through the sieve and fall into the spirit. The advantages of the Berlese funnel are that its size may be adapted to the scale on which the individual is working, and that sacks of leaves can easily be filled in the field and carried back to the laboratory for treatment at leisure.

The pitfall trap is most conveniently a glass jam jar or a plastic cup. This is sunk in the ground, so that its rim is level with the surface. Nocturnal wanderers that fall into the trap find it impossible to climb out of a jam jar while a plastic cup has the advantage that a few holes in the bottom allow the escape of rain water.

Traps intended to provide live animals should be visited daily. Alternatively, the vessel may contain some preservative, such as 2% formalin, when longer intervals may be allowed. Some trials are generally necessary to determine the best siting for the traps: dogs and other animals may disturb them, and the risk of rain is always present. A roof is often a good addition.

For the field or shrub layer the methods of sweeping and beating are appropriate. The sweeping net is a stout canvas bag mounted on an iron ring; it is dragged through the undergrowth or long grass, and its contents periodically turned out on to the sheet. Shrubs, hedges and the lower branches of trees may be shaken or beaten over an inverted umbrella or anything else suitable for arresting the fall of the animals disturbed.

The tree or canopy layer contains the Arachnida that live on or under the bark as well as those among the leaves. It is nearly always rewarding to prise up the bark from the trunk of a fallen or dead tree, but growing trees should not be so ill-treated. Some authorities advise brushing the bark with a soft brush; others, more ambitious, apply the suction of a vacuum machine to the trunk. A good alternative is the wrapping of a band of corrugated cardboard round the tree. This provides the small crevices in which so many invertebrates delight to hide themselves, and when it is unwrapped on the following morning it is seldom found to be untenanted.

To these general methods there may be added the modern and sophisticated method of detecting scorpions by fluorescence. A scorpion's exoskeleton glows under the incidence of ultra-violet light; therefore the scorpion hunter works at night, carrying an ultra-violet lamp with which to survey his surroundings. Scorpions are then revealed by a greenish-yellow glow. Williams, who has developed this device, reported a collection of 2,000 scorpions in 4 hr on a Californian sand dune, adding "At one time seventeen specimens were seen glowing in the light of one lamp".

Forster and Forster (1973) strongly recommend collecting by night, wearing a battery-powered lamp attached to the forehead. It is thus, they say, that one sees spiders behaving in a natural manner. Moreover, spiders' eyes may reflect the light and appear as minute green spots yards away and before the spiders themselves can be seen.

The underlying importance of all the methods described is the fact that the different types of habitat are in general occupied by different species and genera. A scientist therefore takes note of these distinctions when labelling his collection, and adds, moreover, the month or at least the season in which a species occupied the habitat from which it was taken.

PRESERVATION

Arachnida are essentially "spirit-specimens", since their soft bodies shrivel if allowed to dry. In the traditional method they are kept in alcohol, the strength of which should not be less than 70%, in small specimen tubes. A few species will discolour the spirit when first put into it, so that it may need changing. The brilliant colours of some specimens may disappear, but in the majority there is no alteration, and the only problem is the inevitable loss of spirit by evaporation. The addition of 1% of glycerine is an established way of reducing this and it also tends to keep the preserved body supple. After the label with the vital data has been included, the tube is closed with a pellet of cotton wool, or perhaps more satisfactorily with a plastic cap, and is inverted in more spirit in a wide-mouthed stoppered bottle. In these conditions an arachnid can be kept and well preserved for at least 200 years and probably indefinitely.

Good bottles with ground-glass stoppers are an expensive luxury, and may be replaced by screw-topped bottles with plastic tops.

For mere exhibition, as in a museum, the hard specimens, like scorpions and whip scorpions, may be allowed to dry, but as this fixes their limbs in immovable positions it has disadvantages for other purposes. The bodies of scorpions become rigid in alcohol or in alcohol and glycerine, and Williams (1968) recommends an improved formula:

Formalin	12%
90% Alcohol	30%
Water	56%
Acetic acid	2%

This mixture penetrates and preserves the internal organs, especially if the scorpion has been killed by dropping it into boiling water.

Whatever method is adopted, loss of alcohol will occur, albeit slowly, and specimens not regularly inspected will in time be dry, except for a residue of glycerine. This leaves the specimen open to destruction by fungi. There is clearly a need for a better preserving fluid, and one alternative is isopropyl alcohol. Another, well recommended, is propylene phenoxytol in a 2% solution. Specimens may be killed in this; they die with their limbs extended and are then fixed in alcohol and transferred to the phenoxytol.

ANATOMICAL STUDY

For laboratory investigation of the bodies of Arachnida, three methods are available. In ordinary circumstances the animal is most easily examined as it lies in alcohol in a clean white saucer. It may be supported in the desired position by small pieces of broken porcelain or of granulated tin. Fine white sand under spirit also holds the specimen satisfactorily.

A bright, direct illumination is an essential, when the 1-in. objective will be found to give sufficient magnification for all except the smallest specimens. Alternatively, the specimens may be allowed to dry and then examined in the same way but without the alcohol, and in some cases this is found to be an advantage. A third method is unorthodox but is most efficient. The specimen is allowed to dry and is then picked up by one leg with the stage-forceps. The points of the forceps are brought into focus and illuminated. The animal can now be turned about and viewed from all angles more easily than in any other way.

Laboratory examination of the Arachnida also includes dissection and microtomy. Ordinary naked-eye dissection is almost limited to the large spiders, scorpions, etc., which are conveniently embedded in wax for the operation. Reasonable experience in small dissection is necessary, and it is more than usually desirable that one's scalpel blades shall be really sharp.

Section-cutting is more specialized work. The exoskeletons of Arachnida do not allow the rapid entry of fixatives and are hard to cut. In spiders the difficulty is increased by the fluid nature of the abdominal contents and the way in which the eggs become unexpectedly hard. In the study of ovaries, however, these organs can be dissected out and fixed separately. To a limited extent this is also true of silk glands, but to little else.

The trouble caused by the hardness of the chitin has often been avoided by using animals which have just moulted, and whose exoskeletons are still soft; but the process of moulting is accompanied by various internal changes, and, moreover, suitable animals are not easy to obtain.

Millot (1926) has perfected a process by which these obstacles are largely overcome, and his procedure is as follows.

1. Dehydrate.
2. Transfer to a mixture of ethyl ether and absolute alcohol for 12 to 24 hr, according to size.
3. To celloidine solution for 24 hr.
4. To toluene, two changes.
5. Embed in paraffin wax in the usual way.

As a fixative solution Duboscq-Brazil's or Fleming's solutions may be used, but in general the best results are obtained by the use of the following mixture, recommended by Petrunkevitch (1933):

Alcohol, 60%	100 cc
Nitric acid, sp. gr. 1·42	3 cc
Ether	5 cc
Cupric nitrate crystals	2 g
p-Nitrophenol crystals	5 g

Specimens may be left in this fluid as long as desired without the tissues becoming any harder than they are in the living animal. The fixative is washed out with three or four changes of 70% alcohol. The best stains to use are Delafield's haematoxylin, either alone or counterstained tetrabrom-fluoresic acid; or Mallory's triple stain.

Petrunkevitch also recommends the examination of entire specimens, which, after fixing, have been cleared in tetrahydronaphthalene. This makes the muscular and alimentary systems more visible. Locket and Millidge (1951) report improved results in the examination of such external features as epigynes if the animal is lying in phenol to which 10% of alcohol has been added.

Small arachnids like false scorpions, young spiders or young harvestmen can be successfully cleared by soaking in warm lactic acid. They can be mounted in Micrex or alternatively in Faure's medium, made up thus:

Chloral hydrate	10 g
Water	10 cc
Glycerin	2·5 cc
Gum arabic (white, clear)	6 g

This is made cold and is ready for use in a week.

Beechwood creosote and clove oil also clear specimens and epigynes, while Hoyer's solution both clears and mounts pseudoscorpions, Acari and linyphiid spiders adequately.

Lastly, Hopfmann's technique, designed in the first instance for the examination of silk glands in spiders, may be given as follows.

1. The abdomen is fixed in Carnoy's fluid.
2. To methyl benzoate.
3. To methyl benzoate and celloidine.
4. To benzene.
5. Embed in paraffin wax.

In the study and identification of spiders the exact form of the female epigyne and of the male palp is often of the greatest importance. The former may be completely removed, cleaned by a moment's warming in dilute potash and mounted. Preferable to Canada balsam is a synthetic resin, dimethyl hydantoin formaldehyde.

The examination of the male palp is easier, though the organ itself is often more elaborate. It may be expanded by warming in dilute potash and either mounted or, preferably kept in a micro-vial separately from the rest of the corpse, though in the same tube. The removal of the male organ from the abdomen of a harvestman is quite simple; it is chitinous and tough and easily cleaned and mounted as a permanent slide.

BEHAVIOUR

Complete knowledge of the biology of Arachnida cannot, however, be obtained by examination of no more than their dead bodies; their behaviour, their responses to a wide range of stimuli should also be investigated. In this aspect of zoology the Arachnida are often found to be excellent subjects, living with apparent contentment in captivity, and reacting as normally to stimuli as they do in nature. Some species are harder to keep than others, or perhaps it is more accurate to say that the conditions necessary for their survival in cages are less easily discovered and maintained, but this is not the general rule. Most arachnologists have been able to observe one or more species very fully and have described their methods. Millot gives very high praise to the

Thomisidae, which he describes as the ideal laboratory spiders, capable of living in anything. Bonnet's great work on Dolomedes is a model of its kind, unsurpassed in extent and fullness of detail, while in his study of the life-history of Nephila, the same author has succeeded in hastening the development of the females and retarding that of the males, so that members of the same cocoon reached maturity simultaneously, a striking modification of natural events.

In addition to this it is a fact not sufficiently recognized that the Arachnida in general and perhaps spiders in particular provide very suitable subjects for the study of animal behaviour. They are essentially animals of the small-brained type, richly endowed by heredity with instinctive modes of behaviour but without the power to learn that is possessed by big-brained creatures. It is often unnecessary to postulate the existence of a thinking or conscious mind in interpreting their actions, so that Arachnida may successfully be used by mechanistic biologists as examples of automata or organic machines.

It is probably for this reason that they are so ready to live an undisturbed life in our observation cages. On occasions in the past, experiments on insect behaviour have yielded untrustworthy results because the insect did not exhibit its natural reactions when in captivity, and Acari, Solifugae and, sometimes, Opiliones have been found to give untrustworthy results due to behavioural changes when in captivity. But these are on the whole exceptional difficulties. The circumstances of captivity seldom disturb the placid and amenable arachnid, its mechanical responses are called forth and its behaviour can be measured and often predicted, as if one were working with a galvanometer or a spectroscope.

An important feature of the recording of observations made of an active animal is due to the physical limitations of the human observer. The forgetting of a detail may be avoided by immediate note-taking, but it may well happen that by the time the note is made the animal has proceeded to the next stage of its operations, and this may be completed before the hurried observer looks at it again. This can be avoided by those who possess the necessary equipment by making an immediate report of everything seen into a tape-recorder, from which it can be recovered and repeated as often as necessary.

REARING

Bonnet's work on Nephila, mentioned above, as well as that of other competent technicians, affords proof that the life history of an Arachnid can be accurately determined by raising specimens from birth to maturity. The eggs of spiders, harvestmen and false scorpions usually hatch in normal laboratory conditions without difficulty. They run opposing risks from desiccation, which effectively kills them, and damp, which encourages the growth of equally fatal moulds. The general advice to "Keep ova not too warm and not too wet" can only be successfully followed after trial and possible error. Where conditions are controllable a temperature of 20°C and a relative humidity of near 40% are usually near the optima.

After their escape from the egg membrane young spiders must be kept apart in separate vessels; other young arachnids are seldom equally cannibalistic. They need food, and as a rule will accept most of the small invertebrates that normally accompany them in leaf litter. In many laboratories the culture of Drosophila is a continuing operation and provides a readily available source of nourishment. It must be remembered that Arachnida on the whole feed occasionally and digest slowly; also that they usually cease to feed just before ecdysis. Those who deal with exotica, such as scorpions and Solifugae, can replace Drosophila by locusts, which are as easily cultured.

The chief features to be determined are the number of ecdyses that take place between birth and maturity, the length of the intervals between successive moultings, and hence the total life span. These differ from order to order and even from family to family.

An unexpected difficulty is encountered in the rearing of harvestmen, and may, perhaps turn up with other orders. Harvestmen possess the advantage that they do not need to be given living prey, but they counter this by dying an early death after the third or fourth ecdysis. Klee has given both explanation and cure. The cause is the constancy of humidity in the cages, which prevents complete hardening of the new cuticle after moulting. The conditions of nature are more closely followed if the lid of the cage is opened daily to allow the humidity to fall from 100% to about 40%.

Rearing in the laboratory and the consequent production of a complete timetable for the life of a species is one of promising lines of straightforward arachnology at the present time. The cast-off exoskeleton deserves the closest scrutiny, as one by one they display the stages of growth. The admirable work of P. D. Gabbutt on false scorpions and of Max Vachon on spiders of the genus Coelotes are examples of this. The knowledge they have provided often makes it possible to determine the instar of a given specimen from details of its leg joints or its setae.

EXCURSUS

A Laboratory Course in Arachnology

There are probably few, if any, university courses in practical zoology
in which Arachnida can hope to occupy as much as the ten periods for
which the work outlined below is designed. The full course may,
however, be used as a basis from which selections can be made, while a
few specialists or others working privately may wish to follow it completely and to supplement it. In recognition of this, the work has been
divided into two parts, general and special.

GENERAL COURSE. This consists chiefly of examining and drawing the
demonstration specimens provided, and the preparation of microscope slides of entire specimens of some of the smaller Arachnida. The
part occupies five periods of about two hours each.

1. (i) Limulus, dorsal view: Cephalothorax; abdomen; telson; compound eyes; simple eyes.
(ii) Ventral view: Cephalothoracic limbs 1–6, gnathobases on
2–6; digging process on last leg; genital operculum; abdominal
limbs 1–5, with book-gills; anus, telson.
(iii) Casts of fossil forms, such as Eurypterus and Prestwichia.
(iv) Scorpion, dorsal view: Chelicerae; pedipalpi; carapace;
median eyes; lateral eyes; mesosomatic segments 1–7; metasomatic
segments 1–5; telson.
(v) Ventral view: Chelicerae; pedipalpi; legs 1–4 with gnathobases on 1 and 2; genital operculum; pectines; four pairs of stigmata; telson.

2. Araneae
A moderate-sized spider, preserved in 70% alcohol or freshly
killed by dropping into boiling water, is gently boiled in dilute
(10%) potassium hydroxide. The viscera are thus entirely dissolved and the exoskeleton which remains is rapidly upgraded to
absolute alcohol, cleared and mounted in balsam under a large
cover-slip. Under 1-in. objective draw the following.
(i) Dorsal view: Chelicerae; pedipalpi; legs 1–4; cephalothorax;
abdomen.
(ii) Ventral view: Chelicerae; pedipalpi with maxillary gnathobase; labium; sternum; legs; epigyne; book-lungs; spinnerets.
(iii) From a mature male spider the pedipalpi are detached. One is
dehydrated and mounted on its side in its normal state; the other is

boiled in potash until the organ has expanded, when it is dehydrated and mounted.

3. Opiliones

A harvestman with its legs cut through the femora near the body is boiled in potash and mounted whole.

(i) Dorsal view: Chelicerae; pedipalpi; eyes; orifices of odiferous glands; cephalothorax and abdomen broadly joined; vitta.

(ii) The characteristic penis of the male or the ovipositor of the female can easily be extracted from the abdomen after an incision has been made slightly to one side of the middle line. These parts are then mounted.

(iii) After this, the legs of one side of the same specimen are removed. This makes it possible to pull the other group of coxae in the opposite direction, exposing the normally hidden vestigial sternum, and, if the genital plate is pulled back, the first abdominal sternite.

(iv) Permanent mounts of a chelicera, of a pedipalp and of a single tarsus are also well worth making.

4. Pseudoscorpiones

These should be studied by the method described by J. C. Chamberlin. The chelicerae and palpi and the legs of the first and fourth pairs are removed from the body, dehydrated, cleared and mounted. The body is gently boiled in potash, washed, stained with magenta or fuchsin, cleared and mounted. All the characteristic features are well displayed by this method.

5. Acari

Cheese mites, Tyrophagus, and red-spider mites, Tetranychus, are readily obtainable for mounting. A drawing is made from the ventral aspect, showing: thorax and abdomen broadly fused; chelicerae; pedipalpi; legs; genital orifice; anus. Various species of mites are easily found by bringing a boxful of drifted leaves and similar vegetable debris into the laboratory and turning it out on to a sheet of paper. The mites that crawl out are straightway drowned in spirit. Quite satisfactory preparations may often be made by merely allowing their bodies to dry and covering them with a drop of balsam and a cover-slip. But crude methods like this are not necessarily to be encouraged.

SPECIAL COURSE. Owing to the small size of most of the Arachnida, and the difficulty of obtaining the largest species in Britain, this course is of necessity based on the spider. Spiders, however, justify this choice since they are the most highly specialized order. This course also occupies five periods.

6. Families of Spiders

(i) Examine and draw in outline the nine demonstration specimens selected as representatives of the chief European families. The following species are all readily obtainable and are suggested as suitable for the purpose.

FAMILY	SPECIES
1. Gnaphosidae	*Herpyllus blackwalli*
2. Thomisidae	*Xysticus cristatus*
3. Salticidae	*Salticus scenicus*
4. Amaurobiidae	*Amourobius fenestralis*
5. Agelenidae	*Tegenaria derhami*
6. Lycosidae	*Lycosa amentata*
7. Theridiidae	*Steatoda bipunctata*
8. Linyphiidae	*Linyphia triangularis*
9. Araneidae	*Araneus diadematus*

(ii) Examine examples of the Mygalomorphae and Liphistiidae, if available.

7. External Structure

(i) After boiling in potash, the exoskeleton of a spider is dissected and a set of slides made to illustrate all the chief external features. The following are suggested: (a) carapace, with eyes; (b) maxillae with palpi and labium; (c) legs; (d) spinnerets; (e) sternun; (f) chelicerae.

(ii) A leg, parboiled in stronger potash, readily sheds its spines and setae. If then mounted, the lyriform organs are more easily visible.

8. Genitalia

(i) Female: the genital area is cut away from the abdomen, cleaned by gentle digestion in dilute potash, dehydrated, cleared and mounted.

(ii) Male: general treatment as in 2 (iii) above, but different species should be used. For further remarks on the treatment of palpal organs, see T. H. Savory (1927), *J. Quekett Micros. Club.* **xv**, 252–4.

(iii) Spermatozoa: a male spider is chloroformed, the palp removed and crushed between two slides. This produces a smear of spermatoza, together with some debris which is removed. The smear is fixed with Bouin's fluid, after being dried on the slide, and the Bouin then washed off with tap water. Ehrlich's haematoxylin is applied until the smear goes a deep red colour, when it is washed in tap water until it turns blue. It is then dehydrated, cleared and mounted in balsam.

9. Internal Structure

(i) The dissection of the large house spider, *Tegenaria*, is most interesting though not conspicuously easy. With but moderate skill, however, the following should be detectable: in the cephalothorax —poison glands (stain and mount), pharynx endosternite (clean in very dilute potash); in the abdomen—heart, ovaries and testes (stain and mount), one or more kinds of silk gland (double stain with haematoxylin and eosin).

(ii) A spider macerated in water until all the soft parts have rotted away is well worth examination. The pharynx and sucking stomach are more easily seen by this method than by any other and usually remain attached to the lip.

10. Other Arachnida

A more detailed examination of such demonstration specimens as may be available, e.g. Galeodes, Mastigophorus or Thelyphonus and a small type-collection of ticks.

33

Chemical Arachnology

In the study of arachnology there are encountered a number of substances, the chemical constitution of which is of unusual interest. Among these are chitin, silk, venom, hormones and pheromones.

CHITIN

A considerable proportion of the exoskeleton of an arachnid has been believed to consist of chitin, a resistant material that is unaffected by air or water and is insoluble in caustic alkalis. It is dissolved by strong sulphuric or hydrochloric acid, forming a solution from which dilution or neutralization causes it to be deposited in a changed form as a white precipitate.

If into glucose molecule, $CH_2OH(CH.OH)_4CHO$, the amino radical NH_2 is introduced, the product is glucosamine, $CH_2OH(CH.OH)_3 CH.NH_2.CHO$. If next one of the hydrogen atoms of the amino radical is replaced by the acetyl radical, CH_3, the compound so formed is acetyl glycosamine, $CH_2OH.(CHOH)_3CH.NH(CH_2CO).CHO$. Finally, a polymer of this is chitin, to which the following structural formula has been assigned:

When strong acids hydrolyse this molecule, it is decomposed into acetic acid and glucosamine,

$$2C_{15}H_{18}O_{10}N_2 + 6H_2O \rightarrow 4C_6H_{13}O_5N + 3C_2H_4O_2$$

The structural formula displays the presence of the peptide bond, CO.NH, which is characteristic of all protein molecules. Hence chitin has been doubtfully described as a polypeptide. Nor, it may be added, is it synonymous with the comparable substance cuticulin.

SILK

No material is more characteristic of Arachnida than is silk, which is produced by spiders, pseudoscorpions and Acari. Its striking properties and its differences from the familiar silk of *Bombyx mori* have made a determination of its composition desirable, despite the manipulative difficulties involved.

A thread of unspun commercial or "true" silk consists of a core of the protein fibroin covered with a sheath or outer layer of another protein, sericin. A spider's thread has no coat of sericin; the fibroin of which it is almost wholly made is very closely allied to the fibroin of true silk, but is not identical with it. The silk from either animal is stained yellow by zinc chlor-iodide and dissolves in hot caustic potash or in cold concentrated sulphuric acid, but whereas true silk resists the action of chromic acid, acetic acid or ammonia, all these substances cause spider silk to swell or contract.

The first published analysis of spider-silk was due to Fischer (1907). This mentioned nine constituents, most of which were amino acids, of which the chief was alanine at 23·4%.

It was soon realized, however, that the silk from different species is slightly different, as also is the silk from different glands of the same species, and this was included in an analysis by Peakall. This precision was taken a stage further by Anderson (1971), who pointed out that analysis of a whole gland, by including the cells of the gland itself, introduced an avoidable inaccuracy. He therefore analysed the contents of the lumen only. His results mentioned 16 constituents, ten of which were not included by Fischer.

A general impression of whole matter may be gained by comparing the figures of these three analyses, but limited to the chief compounds common to the two most recent sets of figures.

	PEAKALL					ANDERSON				
	Aciniform	Aggregate	Ampullate	Cylindrical	Piriform	Aciniform	Aggregate	Ampullate	Cylindrical	Piriform
Alanine	25·3	27·3	32·7	25·4	29·3	10·7	6·2	17·6	24·5	9·9
Glycine	15·3	20·3	24·3	11·9	24·7	13·2	14·5	37·2	8·63	7·8
Serine	11·7	5·3	6·8	11·7	5·3	16·45	6·8	7·4	27·6	14·8
Argenine	3·8	3·4	3·2	5·0	4·8	4·0	3·4	0·6	1·5	3·6
Aspartic acid	2·1	1·8	1·3	2·3	1·8	8·0	9·2	1·0	6·3	10·5
Isoleucine	4·8	5·7	1·7	5·0	2·1	4·3	4·7	0·6	1·7	3·7
Leucine	3·2	2·9	2·1	3·5	2·2	8·2	5·5	1·3	5·7	5·4
Lycine	2·2	3·2	1·8	1·3	2·1	2·6	7·5	0·5	1·8	9·0
Proline	10·2	12·8	3·1	3·8	4·7	3·6	10·5	15·7	0·6	7·8
Tyrosine	3·7	1·7	1·8	3·7	2·6	1·6	2·2	3·9	0·9	2·2
Valine	4·8	8·5	1·9	2·7	2·9	6·9	5·8	1·1	6·0	5·4

The silk of pseudoscorpions was analysed by Hunt (1970), working with the species *Neobisium maritimum*. His figures make an interesting comparison with those given for spider-silk by Lucas and Rudall (1968):

	Hunt	Lucas and Rudall	
Serine	164	143	(parts per 1,000)
Glutamic acid	103	100	
Glycine	176	248	
Alanine	16	250	
Aspartic acid	53	27	

Pseudoscorpions, like some other Arachnida, also construct or secrete spermatophores, and Hunt and Legg (1971) analysed these, using no fewer than 1,500 examples from the species *Chthonius ischnocheles*. They detected 17 amino acids, chiefly:

Glycine	240	(parts per 1,000)
Alanine	145	
Glutamic acid	123	
Serine	89	
Aspartic acid	75	

VENOM

The composition of the poisons injected by the really formidable Arachnida has been investigated by a number of chemists during recent years, not always with consistent results. The subject has been fully summarized by Junque and Vachon (1968).

There is an overall agreement that some half-dozen compounds are generally to be found. They are:

Spermine, $NH_2(CH_2)_3NH(CH_2)_4NH(CH_3)_3NH_2$
Trimethyldiamine, $NH_2(CH_3)_3NH_2$

The above are present in the venom of most spiders examined. Also in the list are:

Histamine $NH(C_5H_6N)NH_2$
Serotamine or 5-hydroxytryptamine
γ-Aminobutyric acid $C_4H_8(NH_2)COOH$
Para-hydroxyphenylamine $C_6H_4(OH)NH_2$

Mixed with these a protein and an enzyme that decomposes carbohydrates are to be included. It appears that the spermine is the most dangerous constituent of spider poison, while the serotamine is the chief source of danger in the sting of a scorpion.

The various components of arachnoid venom seem to be divisible as to their different effects. One at least has rapid subduing effects on a bitten insect, others act more slowly. When a human suffers seriously, the trouble is due to a component that stimulates the secretion of acetylcholine from the nerve endings, quickly exhausting the supply. Acetylcholine is a neurohormone which causes unstriped muscle to relax. It is destroyed by the enzyme cholin esterase, so that in its absence nervous impulses cannot cross a synapse or reach the muscles they normally serve.

The venom of false scorpions, which is undoubtedly very virulent in its action on the normal prey of these small animals, seems not to have attracted the attention of the biochemists.

PHEROMONES

Pheromones probably play a much larger part in the lives of Arachnida (and other invertebrates) than is readily detected, but recent research had directed attention towards this aspect of animal behaviour, and a number of interesting results have been recorded.

As long ago as 1969, Hegdekar and Dondale showed that sexual behaviour in the wolf spider *Schizocosa crassipalpes* was determined by contact with a sex pheromone. It appears that the compound is secreted by the female, and the suggestion has been made that its origin is in the pyriform glands. Thus the drag-lines of the female are effective when a mature male treads on them, as it seems likely that the chemoreceptive organs are in or near the tarsi. It is postulated that sex pheromones help the sexes to meet, and are especially valuable for species that are not very plentiful in a given area, and where scent and sight would be of less value.

The chemical composition of the production has not been published. However, Berger (1972) isolated the sex pheromone from 50,000 virgin ticks and determined that it was 2,6-dichlorophenol. Similar work with the same number of spiders seems to be indicated as desirable. Dondale and Hegdekar have later (1973) shown that the pheromone is associated with the drag-line of *Pardosa lapidicina*.

Pheromones are also of value in protecting the animals from the attacks of predators. The most readily seen example of this is the use of the odoriferous glands by Opiliones. Estable (1955) analysed the secretion of one of the Laniatores, *Heteropachyloidellus robustus*, and found it to contain several methyl quinones. Two species of the Palpatores were investigated by Blum and Edgar (1971). They were *Leiobunum formosum* and *L. speciosum*, and were found to emit 4-methyl-5-heptanone.

Members of the order Uropygi have long been known to emit a smell of acetic acid when threatened, and often a fluid is squirted from the glands near the base of the flagellum. This was analysed by Eisner (1961), with the result:

Acetic acid	84%
Caprylic acid	5%
Water	11%

HORMONES

The presence of hormones in the bodies of Arachnida may be confidently deduced from their behaviour; precise chemical evidence, however, is rare.

Kühne (1959) described groups of neurosecretory cells in the protocerebrum and chelicerae of Tegenaria, Trochosa and Araneus. He also noted that the acts of ecdysis and reproduction were accompanied by an increased activity of these cells. Legendre (1956) distinguished two differing types of neurosecretory cells in the genus Tegenaria. One type showed up most prominently in males that were wandering in search of a female, and in females just after fertilization. Following this, the cells atrophied in both sexes.

Yoshikura and Takano (1972) described two types of neurosecretory cells in the supra-oesophageal nerve mass and three types in the sub-oesophageal mass in Atypus. In some of these cells seasonal changes were seen, confirming earlier evidence that they play a part in the life history.

Neurosecretory cells in the harvestman *Leiobunum longipes* have been shown by Fowler and Goodnight (1966) to produce 5-hydroxytryptamine in amounts that were greatest at the times of increased activity of the animal; both were at a maximum at 2 a.m. (see Chapter 8).

FLUORESCENCE

There is a substance in the exoskeletons of Arachnida that fluoresces under the incidence of ultra-violet light. It has been seen when a scorpion, Solifuge, Amblypyge or spider is subjected to such radiation. Lawrence (1954) has described how specimens of these orders were irradiated by an 80-W mercury vapour lamp, and that in some the dorsal or ventral sclerites or both showed fluorescence, while in others the effect was confined to the softer intersegmental membranes. The substance responsible for this surprising result has not been named. It appears to be soluble in the preservative fluids used in museums, for some of its first reports mentioned that these liquids also showed the effect.

The phenomenon has been exploited by Williams (1968) for sampling the population of scorpions active during the night, with such success that he has recorded the sighting of 17 individuals in the light of a single beam and the finding of 2,000 specimens in 4-hr collecting. The nature of the compound involved should be discovered. Many compounds are known to fluoresce in this way, among them being chlorophyll, aesculin, paraffin oil, fluorspar, fluorescein, barium and magnesium platinocyanides. The investigations of J. Striano have shown that water becomes fluorescent after the bodies of scorpions have been soaked in it. The fluorescent compound itself can be scraped off the exoskeleton, and is unaffected by exposure to air or to temperatures as high as 100°C. It seems to be secreted by superficial cells of the cuticle, but is also present in the apodemes of the chelicerae. It is possibly a compound similar to the luciferescein of fire-flies.

Arachnologists should be encouraged to increase our knowledge of the chemistry of arachnid physiology, remembering the words of Lavoisier, "La vie, c'est un action chimique".

34

Medical Arachnology

The medical aspects of Arachnology may begin with a reference to Nicander, a physician of the second century B.C., a native of Claros, near Colophon (the home of Arachne herself). He was the author of *Theriaca*, a long poem which included the treatment of a bite from a spider. A few of his hexameters seem to refer to the effects of a Latrodectus bite. His doctoring was recorded by Pliny (A.D. 23–79), advising the application of a cock's brain in vinegar, or a sheep's droppings, also in vinegar.

Spiders themselves were valued by the medical men of the day. The "wolf spider" applied to the forehead in a compress of resin and wax was recommended as a cure for tertian fever. Further, "a spider descending on its own thread, taken in the hollow of the hand, crushed and applied 'appropriately' acts as an emmenagogue, but if the spider is taken as it is climbing upwards it has the opposite effect."

Both Pliny and his contemporary Dioscorides wrote of the value of "cobweb" for stopping the flow of blood from a cut or small scratch. References to this have continued to appear ever since, and the effectiveness of spider-silk in such small emergencies is universally recognized. Its persistence has been remarkable.

In "A Midsummer Night's Dream" Bottom says, "Good Master Cobweb, if I cut my finger, I shall make bold with you" (3.i.90), and the practice was mentioned by both the early British naturalists Edward Topsel and Thomas Moffett (1553–1604). It is to be found in the Diary of Charles Ashmole for 11th April, 1681 and was advised by John Wesley in "Primitive Physic", 1762.

The viability of this trust in cobwebs as a styptic provokes the need to explain its action. On one point all are agreed; that no qualified medical practitioner today would think of prescribing it, but would not deny its value in an emergency. Even now, stablemen are sometimes advised not to clear away the cobwebs in the stable—"You never know when you may need them"—a fact that underlines the surprising absence of recorded bacterial infection from the web. Spiders' webs can hardly be sterile.

A simple and possible explanation is that the action of the threads is no more than a mechanical one, having the effect of favouring the deposition of fibrin. In this case, comparison with the help of a piece of cotton wool is obvious. The "mechanical effect" is most probably an example of the result of distributing unclotted blood over a wide area, which is found to provide a base on which platelets are quickly deposited. This is applied in the modern use of several synthetic materials in sponge form, placed in contact with bleeding surfaces.

Enquiry has, however, carried the subject a stage further. A number of substances of widely differing composition have been shown to function as activators of the Hageman factor, which in turn initiates the clotting mechanism. Among those quoted are kaolin, barium carbonate, asbestos, carboxymethyl cellulose and calcium pyrophosphate, as well as spider-silk. It is not impossible that this is connected with the fact that platelets carry a negative charge, and are deposited electrostatically whenever positively charged surfaces come into contact with them.

The clotting of blood is a complex process, and it seems that the last word on the action of cobwebs or silk has not yet been heard.

Returning now to the eighteenth century, the practice of homoeopathy deserves attention. This system of medicine was introduced by Samuel Hahneman in 1796, and one of its fundamental principles was the use of drugs in very small quantities. Spiders receive special attention. The venom from *Tarentula hispana* is said to influence the nerves supplying the human uterus and ovaries, and to be a remedy for hysterical conditions and similar nervous complaints. *Tarentula cubensis* is similarly described as a cure for abscesses and swellings of any kind. Venom from *Mygale lasiodora* is the basis of a remedy for chorea, but *Araneus diadematus* is more efficacious when the symptoms are aggravated by wet weather.

Few spiders, however, have so fully occupied the attention of the medical profession as has the well known tarantula.

The true tarantula, *Lycosa tarentula*, is widely distributed in southern Europe. Its headquarters, metaphorically and historically, cover an area round the town of Taranto, on the shore of Apulia in south Italy; and it was here, according to one Pietro Matthiole of Sienna, that the first case of tarantism was recorded in 1370.

Tarantism was the name given to the remarkable effects that were alleged to follow the bite of this spider, and which could be cured only by music and dancing. By the end of the fifteenth century tarantism had spread over the whole of Apulia; thence it continued its course, covering all Italy and reaching its height about 1650. By the end of the seventeenth century it was declining, and finally there remained only the tarantella, a graceful Italian dance and the last vestige by which it might be recalled.

The symptoms of tarantism, according to contemporary and later writers, were indeed alarming. The work of Nunez, mentioned below, devoted several pages to describing them, and scarcely an organ or a part of the body seems to have been exempt from the effects of the venom. Pain and swelling, paralysis, nausea and vomiting, lassitude and delirium, palpitation and fainting, continued priapism and shameless exhibitionism are all included, together with acute melancholic depression, ending, unless treatment was provided, in death.

The drugs in use at that time had no effect, and alcohol, even in quantity, did not produce intoxication. The only cure was dancing, prolonged and strenuous, inspired by appropriate music and resulting in copious sweating.

A phenomenon of this nature was bound to excite interest, and for several decades descriptions of authenticated cases of tarantism alternated with equally confident denials that the attacks were anything more than midsummer madness or general hysteria. The first writer to gather these together and to offer a reasoned discussion of the disease was Ferdinand Epiphane of Messina, but a more ambitious treatise was that of Baglivi.

Georges Baglivi was a well known Italian physician and his work, "De Anatomia, Morsu et Effectibus Tarantulae", was published in 1696. It occupies only 42 pages and contains what appears to be the first published drawing of the tarantula spider itself. Baglivi described all the symptoms, credible and incredible, various methods of treatment, notes the greater frequency of the disease in July, and discusses various methods of treatment, including music. His work, though short, is of interest since it is the first discussion of the matter by a writer of scientific experience and of a more than local reputation.

He believed that tarentism was a true clinical entity resulting from the bite of a particular spider. This was the opinion of the time, and in Britain the Hon. Robert Boyle in 1686 wrote that his own doubts had been replaced by conviction.

Fifty years later Dr Richard Mead contributed to Eleazar Albin's "Natural History of Spiders" a four-page section "Of the Tarantula", and in his later writings expressed himself at greater length. He wrote that in winter the bite of the spider is harmless, but that in the dog-days the patient is soon "seized by a violent sickness, difficulty of breathing and universal faintness being asked what the ail is makes no reply, or with a querulous voice and melancholy look". This condition is not "relieved by the usual alexipharmic and cordial medicines" and "music alone performs the cure".

"They dance", says Mead, "three or four hours, then rest. At this sport they usually spend twelve hours a day, and it continues three or four days; by which time they are generally freed from all their symptoms, which do nevertheless attack 'em about the same time the next year". Mead also reported that the tarantati, while dancing "talk and act obscenely and rudely, take great pleasure in playing with vine leaves, with naked swords and red cloths" and "cannot bear the sight of anything black".

Disbelief in the reality of tarantism was first expressed by N. Fairfax in the 'Philosophical Transactions" of the Royal Society in 1667; and the opinion was more forcibly expressed in 1671, that is 24 years earlier than Baglivi's treatise, in a letter from a Neapolitan physician Thomas Cornelius. He said that the tarantati were either malingerers or they were wanton young women, described as *dolci di sale*, or half-wits. The case against tarantism was fully argued by Dr Francesco Serao in 1742. His book, "Della Tarantola osia Falangio di Paglia", was published in Naples and contained 284 pages.

The conclusion is, however, unavoidable, that affliction with tarantism must have brought some advantage to the victims, for the phenomenon lasted for a long time, and having, as one might say, exploited Italy, to the full, it reappeared in Spain.

The first published account of Spanish tarantism was a record of six cases described in a "Tratado del Tarantismo" by Dr D. Manuel Iraneta y Jauregut. This was in 1785, or 47 years after Seras's denial and 89 years later than Baglivi. Iraneta's work of 121 pages was, however, but a trifle when compared with that of Francisco Xavier Cid, "Tarantismo Observadoin Espana", published in Madrid in 1787.

Like Baglivi, Cid was a physician of some eminence. He wrote an authoritative account of the disease, based on 38 cases which, he satisfied himself, were well authenticated. He showed that the Spanish and Italian forms were identical, and he accepted as proven the reality of tarantism and the curative power of music. Indeed, he ended an interesting and erudite work with a chapter on "The Philosophy of Music".

Cid's book inspired a number of Spanish doctors to write on the same subject; among them may be mentioned Mestre y Marzal, 1843; Mendez Alvaro, 1846 and Ch. Ozanan, 1856.

The most comprehensive work on tarantism is J. Nunez's "Etude Medicale sur le Venin de la Tarantule", published in Paris in 1866. A book of 168 pages, it begins with a historical review of the records of the complaint, and emphazise three points—that tarantism is a real disease, that music is the only cure, and that it shows a tendency to reappear. This is followed by an account of the spider itself, and a very full description of the symptoms and treatment of the effects of its bite.

The really surprising thing about tarantism is that a quite genuine belief in its relation to a harmless spider should have persisted for so long and until so recent a date as the middle of the nineteenth century. As we have seen, denials of its validity began as early as 1667, but more interesting than denial is an explanation of what is really a remarkable phenomenon.

It appears that the first writer to offer such an explanation was the anonymous editor of an English version of an extensive "Natural History" by Jo Frid Gmelin of Gottingen. This was published in 1795 and contains the following paragraph:

> "The patients are dressed in white, with red, green, or yellow ribbons, these being their favourite colours; on their shoulders they cast a white scarf, let their hair fall loose about their ears, and throw their heads as far back as possible. They are exact copies of the ancient priestesses of Bacchus. The orgies of that god were no doubt performed with energy and enthusiasm by the lively inhabitants of this warm climate. The introduction of Christianity abolished all public exhibition of their heathenish rites, and the women durst no longer act a frantic part in the character of Bacchantes. Unwilling to give up so darling an amusement, they devised other pretenses. Accident may have led them to the discovery of the tarantula; and upon the strength of its poison, the Puglian dames still enjoy their old dance, though time has effaced the memory of its ancient name and institution."

Successors to the tarantula are undoubtedly the black widow spider, *Latrodectus mactans*, some species of the genera Loxosceles and Chiracanthium and a dozen or so scorpions of the family Buthidae.

The symptoms of Latrodectus bite are alarming and widespread. Beginning as a slightly flushed swelling at the site of the bite, pain is soon felt in the armpits and groin. This grows in severity as it spreads to the thighs and torso. There is nausea, followed by cramp, copious sweating, difficulty in breathing and irregularity of heartbeat. The symptoms gradually diminish and may disappear in three days, but in severe cases they may persist for a fortnight.

The bite of the six-eyed *Loxosceles reclusa* is followed by conditions which, first described in 1872, were traced to the spider only in 1957. Unlike the bites of Latrodectus, the effects are localized, beginning with a blister at the site of the wound. The skin turns purple and black over an area of three to four square inches, and as it falls away there is left a pit that slowly fills with scar tissue.

The differences between the two bites can be shortly described by saying that the former is neurotoxic and the latter is haemolytic.

A spider that has been described as the most dangerous European species after Latrodectus is *Chiracanthium punctorium*. It is inclined to enter houses in the autumn, and Maretic (1975) has described the consequences of its bite. There is severe pain, reddening and small local necroses near the wound, followed by enlargement and tenderness of the regional lymphatic nodes. He adds that "chiracanthism" certainly occurs in many countries, but is not recognized as such by physicians who are not familiar with the symptoms.

The sting of a dangerous scorpion resembles both in effects and symptoms the bite of a Latrodectus. The patient's temperature rises and falls as in fever, he may become delirious and nausea may provoke vomiting. It is desirable to check the spread of the poison by the use of a tourniquet where this is possible. Pain may be relieved by barbiturates or by the injection of calcium glucosate.

During recent years much research has been directed towards the preparation of effective antitoxins or antivenins. Their use has been most successful in cases of spider bites. Against scorpion stings a difficulty is introduced by the fact that the differences between the species of scorpion seem to necessitate different antivenins, and identification of the attacker by its victim is seldom either possible or reliable.

One of the most widespread troubles that are traceable to an arachnid is "scrub typhus", endemic in east Asia. This disease is due to Rickettsia, transmitted by the bites of the trombiculid mites close relatives of the "harvest-bug" of this country, and normally an ectoparasite of the native rats. It was well known to our troops in India; in 1934 there were reports of 108 cases, yet in Burma and Ceylon it was, at that time, almost a medical curiosity. The disease became much more serious during the Second World War, and from 1941 the arrival of large numbers of soldiers in the jungle and other sparsely populated areas gave the mites the opportunity to feed on man as well as on rats. The result was that scrub typhus became second only to malaria as a feature of the campaign; in 1944 there were 5,000 cases, of which 350 were fatal.

The problem was attacked in the field and in the laboratory; a Scrub Typhus Research Unit was instituted, with headquarters at Kuala Lumpur, and by the end of the War the disease was under control. The antibiotic chloromycetin clears up the trouble in a day or two, and as a protective measure clothing may be treated with an acaricide such as dibutyl phthalate.

A new aspect of medical arachnology promises to develop from recent researches at Montpellier under the guidance of R. Legendre. Scorpions of the species *Buthus occitanus* and spiders of the genus Argyrodes have been found to be infected with Rickettsiella, apparently of an undescribed form.

35
Linguistic Arachnology

To those who have found the fascination of words, arachnology offers an unexpected treat.

In Old English the word *spinnen* described the work with warp and woof in the familiar domestic occupation, and so gave rise to the noun *spinster*. A masculine form of this was *spinthron*, which in Middle English became *spinthre*, *spither* and so, in due time, *spider*.

A spinster being essentially human, the small eight-legged spinner was first known as an *atorkoppe*, derivable from *ator*, poison, and *copp*, head. This recalls not only the German *Kopf* but also the Welsh *pre-coppin*, the Danish *Edderkop* and the Polish *pajakov*. In Middle English *coppeweb* was the spelling of the modern cobweb.

In Wyclif's translation of the Vulgate, 1382, the word used was *spither*, with an alternative *areyne* or *aryene*. "Arain" was in use in the north of England until quite recent times, and is possibly still to be heard.

Obviously, *areyne* must owe its existence to the Latin *araneus*, used by Lucretius, and to its feminine form *aranea*, used later by Ovid. From these came *araneolus*, Virgil, and *araneola*, Cicero, for a small spider; as well as *araneosus*, Catullus, full or webs.

From the same classical source come the French *araignée*, the Spanish *araña*, the Portuguese *aranha* and the Italian *ragno* in a series contrasting with the German *Spinne* and the Dutch *spin*.

Of more recent years, it can but be stated that the maltreatment of the Latin language by writers about spiders passes belief. A few examples may be mentioned.

Simon's division into Araneae theraphosae and Araneae verae was classically sound, but comment cannot be withheld from all authors who, without exception known to me, have written about a sub-order which they have called "true spiders". Now the contrast to this would be "false spiders", a description that cannot possibly be applied to the Theraphosomorphae. No one seemed to have paused to consider what

Simon meant by "true", or to have realized that he was too sound a scholar to have written anything so foolish. If his successors had remembered the Latin that they learnt at school, or, in the absence of such a luxury, had looked at a dictionary, they would have found that Araneae verae should properly be translated as "ordinary spiders". By thus contrasting the familiar spiders of the countryside with the four-lunged diggers of silk-lined burrows, Simon was making a clear and sensible distinction.

In 1892 Pocock gave the Araneae verae the name of Arachnomorphae. This was almost invariably used until 1933, when it was criticized on the grounds that it might lead to confusion with the name of the class. It was displaced by Araneomorphae, which empirically is as likely to be confused with the name of the genus Araneus. It is also one of the obnoxious hybrids that too often come into misshapen existence, and should never have been allowed to appear in print. Arachnologists, like other scientists, are often compelled to make new words, and they should be encouraged to make them well, not badly. There is no merit in deliberate malpractice.

Some authors have taken to writing "Mygalomorpha" and "Arachnomorpha". Now *morphe* makes the plural *morphai*, which can only be transliterated into *-morphae*. What mental kink suggested this change is obscure.

A similar step has been taken by those who have written "Cribellata" and "Ecribellata", saying that these spellings are "preferable" to the usual Cribellatae and Ecribellatae. No reason for the preference is given. The words can, of course, only be construed as adjectives, qualifying the noun *araneae*, understood. Araneae cribellatae is analogous to Simon's Araneae verae. Long-standing custom has justified the dropping of the noun: we speak, for example, of gastritis, arthritis and so on, when these words were originally adjectives, qualifying *nosos*, a disease. Gastritis nosos meant a disease affecting the stomach. To put these words into the form suggested is pointless, or worse, since *Cribellata*, by itself, can only mean either "a small-sieved woman" or "small-sieved things", and Mygalomorpha means nothing, except that the writer, like Shakespeare, had little Latin and less Greek.

The cribellum seems to have been devised by a malicious evolutionary jinx to give trouble to arachnologists. The organ is to be described as a small sieve, and the Latin for sieve is *cribrum*. Diminutives in Latin are made by adding the suffix *-ulus*, as in puerulus, a little boy, or *-cellus*, as in navicella, a little ship. This *cribrum* might give either *cribulum* or *cribrellum*, and certain authors have, in fact, used the latter form. There is, however, sufficient precedence for the dropping of the second *r*, most familiarly in the word *cerebellum*, a little brain, from *cerebrum*, a brain. I have never seen the word *cerebrellum*, and do not think that *cribrellum* is to be encouraged.

The spinneret, metaphorically described as thele, a teat, produces another curiosity. There are two genera of spiders whose names indicate that they possess six and seven spinnerets respectively. One of these has been named Hexathele and the other Heptathela. The former is an exact transliteration of the Greek, but leaves the word in the singular, "six-spinneret". The latter has converted the Greek into the latinized form *thela*, and so has produced a hybrid that might better have been written Septemthela.

A word that suffers from uncertainty in the minds of its users is *sensillus*. It may be found with all possible spellings as sensillus, sensilla and sensillum. In one standard textbook it appears in the text as sensillum with its plural sensilla, which in the legend to the text figure becomes sensillae (!)

The word seems to be a diminutive from *sensus*. The suffix *-illus* is an alternative to *-ulus* or to *-ellus*, rather less commonly used; but compare *putus*, which makes *putillus*, a little boy. It follows that the correct word is *sensillus*, plural *sensilli*; other forms, including "sensor" are to be avoided.

The matter of inconsistency or of unexplained changes has been mentioned by Kaston (1974). He instances the names that Petrunkevitch gave to some of the higher taxa: the Dionycha and Trionycha of 1928 became Dionychae and Trionychae in 1933 and again Dionychi and Trionychi in 1955. Similarly, the Hypochilomorphae and Quadrostiatae of 1933 appeared in 1958 as Hypochilina and Quadrostiati. Kaston concludes, "I am unable to suggest an explanation of all these changes".

The preceding paragraphs seem to justify the sneer that since scientists are traditionally supposed to be unable to write good English, it is unrealistic to expect them to write good Latin, or even to write Latin at all if they can manage to avoid it. This is achieved by the habit of anglicizing the names of taxa by dropping the final syllable. Thus we read of "liphistiid spiders", spelt with a small *l* or of "galeodids", "gongyleptids" and so on, throughout the whole of arachnid taxonomy. The device may be meant to ease the acceptance of strange names: perhaps it does so.

The high reputation of a science such as ours deserves, in every respect nothing but the best. Are we not arachnologists?

Are we? "Arachnology" must, from its derivation, be the science or study of spiders: the word cannot accurately be applied to students of scorpions, harvestmen or any of the other orders. The comprehensive science should be "arachnidology", a word which, I am glad to say, I have seen in print only once. Few of us are arachnidologists; most of us wisely limit ourselves to the one order that particularly fascinates us.

For more than 30 years I have been accustomed to write of "opilion-ology" and to call myself an "opilionologist", words that are to be recommended on account of their euphony. And in 1966 I coined the "araneist" (pronounced ara-náy-ist, not araníst), to denote one who studies spiders only and neglects the rest.

It would be interesting if the students of harvestmen called them-selves "phalangists" (though this is a *nom. praeocc.*) ; while the Amblypygi were followed by "phrynologists" and the Uropygi by "thelyphon-ists". Specialists in the Palpigradi could rejoice in the name "micro-thelyphonologists". Arachnologists need not be ashamed to confess, with Oliver Edwards, that"cheerfulness was always breaking in".

A subject that is not unconnected with the above is that of the pro-nunciation of the many names that arachnologists have composed in their taxonomic progress.

Half-a-century ago arachnologists were so rare and so widely dis-persed that their communications with one another were almost wholly written or printed: the sounds of their speech were immaterial and were seldom considered. Today circumstances are luckily very different, and it is a common experience to hear a name pronounced not necessarily wrongly, but unexpectedly. Generally the surprising sound can be quickly interpreted: to hear "Nem-ástoma" is at once to recognize "Néma-stóma" but sometimes, and especially when speaker and listener belong to different nationalities, the characteristic intonations of either may delay understanding for a while.

The British Arachnological Society raised the matter in 1971, and produced the following suggestions:

1. In names of two syllables, the stress or accent falls on the first syllable: e.g. Zóra, Phlégra.

2. In names of three or four syllables, the convention of stressing the antepenultimate syllable is generally defensible: e.g. Sítticus.

3. But if the penultimate is naturally an accented vowel, the stress falls on it: e.g. Lophómma, Cyclósa.

4. The antepenultimate tradition is unacceptable for names com-pounded from two derivatives, which should be kept separate: e.g. Dolo-medes, not Dol-omm-edes, but some authorities say that junction is permissible on the grounds of convenience.

5. The *g* in gnathos should be silent: e.g. Tetra-natha, not Tetrag-natha. A Greek gamma is always hard, hence Teggenaria, not Ted-genaria.

6. Names of five or six syllables often fit the antepenultimate rule, but usually acquire a second stress near the beginning: e.g. Pépononcránium, Óreonétides.

Appeal was made to two professors of phonetics for guidance, and they agreed that no precise rules could be considered to be universal. Their advice was that each group of specialists had the responsibility of coming to agreement among themselves.

Just as there have been some who have said that the chemistry of chitin, silk and venom is not arachnology, so there must be many who will feel that philology has no part in the study of the small animals that attract us so strongly. Their sentiments will be echoed by those who criticize zoological nomenclature, saying that zoology is not the study of animals' names. But just as we interpret the behaviour of our contemporaries by saying that it takes all sorts to make a world, so also may it be claimed that it takes all sorts to make a complete arachnology. Or we may change the words of Terence, while preserving his sentiment: "Arachnologicus sum; arachnologiae nil me alienum puto", a rough translation of which is "To be scientifically-minded it is neither necessary nor advantageous to be also illiterate".

V. HETEROGRAPHIA ARACHNOLOGICA

36

The Spider's Web

The webs that spiders spin are objects of an exceptional interest which they unquestionably deserve. They are the chief among the few things that invertebrate animals build or construct, and they are made of silk, in itself an extraordinary substance. Beavers build dams, birds make nests, bees make combs, termites build cities, but only the spider constructs a trap. Silk is produced by false scorpions and mites and by some insect larvae, but they do so for occasional uses only, whereas the spider depends throughout its life on constant silk-secretion. What Tilquin calls "sériciphilie" is its outstanding characteristic. Probably, too, webs have attracted attention and have acted so successfully as advertisements that much of man's original interest in spiders, and hence in other Arachnida, can be traced back to his seeing spiders' webs. In other words, were it not for their webs, spiders would be of as little concern as are silverfish or centipedes.

These basic facts raise many problems. How did web-making originate? How have the different kinds of webs come into existence? Why does each different web-pattern belong to a different family of spiders? How does a spider find its way across its web, and how does it make its way back? And so on. There is no end to such questions, and the best known of all webs, the geometric or orb-web, is the most baffling of them all.

The earliest of all spiders cannot be supposed to have been a web-spinner: probably, like other Arachnida, it was a wanderer, devouring what it was able to catch, but, because it was a spider, trailing a thread of silk behind it. To this it is reasonable to add the habit, widespread among all cryptozoic animals today, of coming to rest in some crevice or hiding-place, which gave it protection and shelter.

From here short sallies to pounce upon passing unfortunates would be an obvious way of life, and a return to safety would be guided by the action of the lyriform organs. Thus there began the use of silk to help in the securing of food, for repetition of this habit would have had an obvious result. The drag-lines would not only form a silk lining to the shelter, they would also radiate from it in all directions, as if the lining had been combed out to form a bell-shaped fringe.

These diverging threads would acquire an unexpected value. Passing creatures would trip over them, causing them to pluck at the silk mat on which the spider was at rest, and the slight movement underfoot would stimulate it to rush out and secure a meal. Thus the stumble-threads, as they have been well called, would become a permanent feature of the spider's neighbourhood.

This account may be better described in different terms. That spiders pounce upon an insect as soon as it has been seen or touched may be regarded as a reflex action, to which the sight or the touch of the victim was the unconditioned stimulus. In the circumstances outlined above the trembling of the silk precedes the leap on the prey, so that the vibrations become the conditioned stimulus, setting in motion a conditioned reflex and so establishing the type of behaviour that is common to all web-spinning spiders.

The structure which may be called the proto-web is found in many families today; it is much more widely distributed than any other design, and it is evident that its first makers had hit upon a very successful mode of life. They flourished and they spread into different types of environment. While some retained their insidiatorial habits, others climbed rock faces or tree trunks; some climbed into the stems and leaves of small plants, some took to running or leaping, some went to the neighbourhood of water or even the sea, some burrowed or dug into the ground. This account of exploratory activity is not peculiar to spiders. The sentences just written might equally well have been written about mammals or any other group of successful animals. The phenomenon is generally described as adaptative radiation, and is accepted as a useful concept by all phylogenists. It is completely applicable to a comparative study of spiders' webs.

This suggests that the different forms which webs show may be regarded as due to two chief causes, namely the nature of the environment and the structure of the spinner.

The rock-face spiders, with which may be included all those that took to sloping ground, show the first modification. In such a situation there is an obvious tendency to run in a downward direction rather than in an upward direction, and to run farther downhill than uphill. As a result the fringe or bell-mouth was no longer symmetrical; it spread downward over an increasing area, producing a primitive type of the familiar cobweb. This is in essence a silk tube, from the lower edge of which a sheet or hammock stretches outwards. Webs of this type are found in several families.

Spiders which climbed into the undergrowth escaped from their enemies on the ground and found shelter under growing leaves, but as they spun their drag-lines from stem to stem they must have found difficulty in maintaining a foothold on the silk. They were not tightrope walkers; they turned over and hung upon their threads, with the help of their toothed claws to support them.

Thus there arose an important new type, the inverted spider beneath its threads, in contrast to the erect spider on its sheet. This is the web characteristic of the hedges and bushes, the web that is associated with the family Linyphiidae.

Two other types of web may be regarded as derived from it. The scraps of silk sheet, with a minimum of threads above or below are characteristic of the related family, the Erigonidae. They have come down from the shrubs to live near, on or even under the earth, among the innumerable cryptozoa, which must have provided them with much of their food.

Some of the plant-dwellers climbed higher. They found themselves among flowers, and in a brighter and more colourful world, and they adopted colours where other spiders went black. Their webs are of interest because many of them seem to present no design, but to be merely a haphazard tangle of threads, spun fortuitously in any direction.

That this is not the truth has been shown by the description given by Lamoral (1968) of the system underlying the making of the web of *Steatoda lepida*, a contribution to the study of webs of the highest significance. Starting from its refuge, the spider spins a few radiating threads attached to convenient, neighbouring supports. These threads are called by Lamoral the primary leading threads. To them are added a number of secondary leading threads, also starting from the refuge or near it. All these are then joined by short primary interconnecting threads, from which in turn secondary and tertiary connecting threads may rise.

By colouring the leading threads at the time of their appearance Lamoral showed that they preserve an unbroken individuality throughout the finished tangle. The whole is not a tangle, but a co-ordinated system, through which vibrations can be efficiently transmitted.

This must be followed by the observations made by Szlep (1966a) of the web-spinning of *Titanoeca albomaculata*, a cribellate spider. From its resting-place the spider spins about four "supporting threads"; others are then added and the typical hackled bands of cribellate spiders are laid on them as "the spider moves like a pendulum, back and forth". Here is a close parallel between a cribellate and an ecribellate web.

Further, webs of Latrodectus may be spun among long grass by a method that closely resembles that of Steatoda. They possess, however, a number of vertical threads, which run downwards to the ground and at their lower ends have viscid drops that delay crawling insects.

These viscid drops place the theridiid web in the position of being the precursor of the orb-web. Were the leading threads to be increased in number until they surrounded the refuge, and were they then joined by intercommunicating threads, the basic design of the orb-web has sprung into existence.

Whether this is so or not, the fact that Latrodectus webs spun in the grass are based on the same principle brings into consideration two other examples of web design in the Agelenidae and Uroboridae.

Anyone who has kept our familiar Tegenaria in a box knows that the spider first establishes itself in one corner of its cage, and during the first evening draws out a small number (four or six) of single, unconnected, radiating threads. These are attached to distant spots on the box sides. Later, treading on these threads, the spider connects them with cross-threads, using the swaying, side-to-side movements so characteristic of the family. As this behaviour is nightly repeated the web grows in thickness and strength though not appreciably in area.

Lamoral (1970 *in litteris*) has drawn my attention to an Uloborus found in South Africa which spins a calamistrated, funnel-shaped orb-web. This is shown diagrammatically in Fig. 107.

This, we believe, represents the primitive type of orb-web. Its relation to more advanced webs is shown by the fact that though in these webs the shelter or refuge is some distance from the web itself, the signal thread is still connected to the central hub. A point of some importance is that the principle is demonstrated equally by calamistrated and non-calamistrated webs, and so strengthens the modern belief that the presence or absence of a cribellum is an almost insignificant detail.

An overall view of these four webs, namely those of Latrodectus, Titanoeca, Tegenaria and Uloborus, give strong support to the hypothesis that all webs grow from the same fundamental course of movements. Further consideration should be given to the application of this hypothesis to other web patterns, for example those of Hyptiotes and Alloepeira.

The orb-web will always be the one that provokes the most baffling problems in web study. Its beauty is not a biological matter, but its symmetry is, for in its apparent symmetry and its deviations from perfect regularity must lie the clues to its construction.

There may be recognized certain stages of web-spinning, which merge into a continuous process, and we must say that instinctive actions are responsible for their orderly succession. We believe that to some extent the directions in which the spider moves are determined

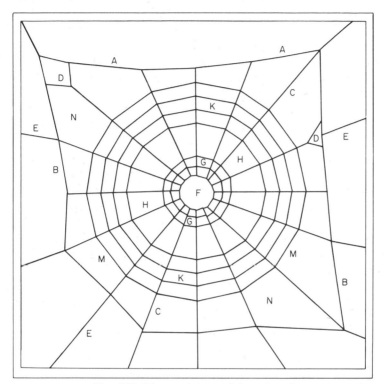

FIG. 107. Nomenclature of the orb web.
A, Bridge thread; B, frame thread; C, radius; D, divided radius; E, mooring thread;
F, hub; G, strengthening zone; H, free zone; K, capturing zone; M, viscid spiral;
N, sector.

by the tensions of the threads on which it treads, and evidence of this kind builds up to the conception of the web as a gradually developing field of forces, to which the instinctively driven spider is compulsorily obedient. One thing appears to be certain; that when once the spinning has begun it continues in a self-determining manner. Tilquin (1942) suggested that the form of any one orb-web is dependent on a few initial proportions. He depicts the web as a dynamic whole, in whose growth physical influences play a dominant part in guiding the responding spider.

Tilquin's contribution has been followed by much careful investigation of web problems, carried out by Witt, Reed, Peakall and others. The Annual Meeting of the American Association for the Advancement of Science included a symposium on web-building spiders; this work has continued and results have owed much to improved photographic methods and the use of a computer in analysis of the many measurements that have been recorded.

The following is a selection from among the many conclusions that have been reached in attempts to improve our understanding of the art of web-spinning as practised by Araneidae.

The electron microscope shows that the spiral thread of an uloborid web is composed of two strands, each 0·5 μm in diameter. There is, in orb-spinners only, a well-defined valve, furnished with muscles, which determines the diameter of a thread. There is close resemblance between this apparatus in Uloboridae and Araneidae, but detailed differences show that all the muscles cannot be homologous. Convergent evolution is suggested. There is definite regulation of silk synthesis within the ampullaceal glands, and the amount of silk remaining in the lumen of the gland, and therefore available for immediate use, affects the type of web produced, probably by a feedback mechanism.

There are such differences between the webs of individual spiders that an experienced observer can recognize the webs of a particular animal, or of one from a particular cocoon, as well as one of a particular species. The webs of young spiders differ from those of mature ones. If a young nymph is taken from the cocoon or egg-sac immediately after it has hatched, it will at first produce only an irregular tangle of threads but if allowed to remain in the egg-sac until the normal time of its emergence "definite maturing processes" have occurred, and the first webs are of regular form though smaller in size (H.M. Peters).

The effects of a laser beam on the central nervous ganglion showed by the spinning of webs with fewer radii or fewer spirals. The effects of many drugs, taken orally when the spider drank sugar solution, have been described. For example, amphetamine caused irregular spacing of radii and spiral; scopolamine produced a wider deviation in spiralling, and LSD resulted in unusually irregular webs. Other drugs affected the size of the catching area and the length and thickness of the threads.

Many experiments by P. J. Peters in transferring spiders from one web to another during the act of spinning showed the importance of external stimuli from existing threads. Similar effects have been seen to follow from such stimuli as the tensions of various threads and the angle between adjacent radii.

Since work on the problems of orb-web construction is likely to continue, an agreed nomenclature for its different parts is clearly desirable. In Fig. 107 the terms suggested by Jackson (1973) and Savory (1952) are combined.

37

Courtship in Arachnida

The existence of courtship among invertebrates is a phenomenon that requires explanation. Among Arachnida it is widespread and is often complex, so that there must be some essential which established the habit in the first place, and some advantage which has favoured its retention and development.

It is well known that the courtship of spiders was first observed in detail by Dr and Mrs Peckham of Milwaukee as long ago as 1889. The spiders they were watching were Salticidae, a specialized family characterized by large eyes, keen vision and bright colours. The vivid description which these authors wrote attracted attention; their conclusion, which was based on the theory of sexual selection, was unfortunate, if, at that time, inevitable. After a number of more or less isolated descriptions of courtship by various writers, the subject was again studied by Montgomery in Philadelphia, by Gerhardt in Germany, and by Bristowe and Locket in Britain.

The general conclusion that followed from the observations of most zoologists was that courtship was a method of conveying to the female spider the information that a male was in her neighbourhood, with the corollary that he was not to be treated as an intruder, killed and eaten. As long as the courtship of spiders only is considered, the theory of sex recognition can be maintained, but when other Arachnida are compared with them, different opinions are possible and the courtship of spiders is seen in a new light.

By starting with spiders, the theoretical study of the behaviour gets off on the wrong foot. There are four reasons for this.

1. The observer of spiders is watching the phenomenon at the top of the arachnid evolutionary scale. Historically, it was an unlucky chance that sex behaviour is very easily seen in spiders and it differs considerably from that of the other orders.

2. The interpretation has been disproportionately influenced by a belief that the female "always" kills and eats the male, who was

dramatically pictured as occupying a terrifying position between fear and desire, seeking, be it only temporarily, the appeasement of his formidable partner. This cast a false light on male behaviour and writers tended to forget that many cases of slaughter had taken place in captivity, when the male was unable to escape; that there had been very few reports of this tragedy from observations in the field; that often the sexes live amicably together for a while; and that sometimes the report told of the male eating the female. The truth is that there is not, and there never has been, any statistical support for this myth of the cannibal female. It probably arose from the chance that the most commonly noticed of all spiders, *Araneus diadematus*, is one of the few species which behave in this way.

3. Courtship is not confined to Arachnida, though they have developed it as least as fully as has any other class. It is to be seen among Mollusca, Insecta and Annelida; and an "explanation" based wholly on the behaviour of a few spiders cannot give a satisfactory interpretation of the phenomenon in general.

4. The elaborate courtship of spiders must be the result of a long evolutionary process. This cannot have been a consequence of a necessity for protecting the males because, in the earliest days before courtship had become established, the males must, *ex hypothesi*, have been killed. There would then have been no offspring, and no race of spiders would have survived to develop the courtship as we see it now.

Continuing the historical treatment, we come to the descriptions of courtship in false scorpions. First described by Kew in 1912, it was little heard of until the researches of Chamberlin in 1931 and Vachon in 1935. False scorpions introduce us to a new process, insemination by means of a spermatophore. This device is a product of the male system, composed of a material that hardens on or after secretion; it supports a droplet or a small sac of semen in a suitable position. A spermatophore may be regarded as a substitute for a penis, much as a spider's web is a substitute for such sense organs as would lead the predator to its prey. The introduction of a spermatophore as an accessory in the reproductive process makes the mating of spiders essentially different from that of false scorpions, and thus divides any discussion of courtship into attempts to answer two questions: how is the courtship conducted, and, why in any case is it undertaken?

Answers to the first question are obtainable by patient observation, which reveals the diversity of the methods adopted by Arachnida of different orders, and even of different families within the same order. Spiders are the most conspicuous in this respect, and instead of trying to give descriptions of a chosen number of types, we shall make use of the evolutionary survey published by Platnick in 1971.

He was primarily concerned with the different natures of the stimuli to which the male spider reacted by beginning the actions that form the main part of the courtship. Discussing the behaviour from an evolutionary point of view, Platnick divides courtship into three grades or levels. He adopts Tinbergen's principle and thus regards courtship as acting as a releaser, which induces corresponding responses in the individual courted, and postulates:

At Level 1: the prime releaser is direct contact between the male and female.
At Level 2: the prime releaser is chemotactic action of silk or a pheromone.
At Level 3: the prime releaser is the sight of the female.

The families in the first group include most of the Therophosomorphae, and the Dysderoidea, Lycosoidea, Aranoidea, Thomisoidea and Clubionoidea. When two spiders touch each other there is created a situation in which one of two consequences may follow. If they are of the same sex or of different species, aggression may appear and fighting may result. If they are male and female of the same species, there is little delay before the insemination of the latter is accomplished. Sometimes, however, the male may grasp the female's leg in his chelicerae, or spin a few threads over her, or even tie her to the ground.

In the second group are many of the families of spiders that spin webs, Theridiidae, Linyphiidae, Araneidae and others. Pheromones have probably played a part in bringing the wandering male to the web; without some such guidance along, fortuitous wandering might be beyond his strength. Arrived at the edge of the web, the male reacts typically by drumming upon it, twitching it, cutting a hole in it or practising stridulation. It seems reasonable to deduce that this has the effect of inhibiting the normal aggression of the female towards any other disturbance of her web. In due course the male advances and physical contact is followed by insemination.

Into the third group fall the keen-eyed Salticidae and Oxyopidae, in which families the males "dance", sometimes for a long period, before the watching females. Often epigamic colourings on legs or palpi are displayed and seem to help to stimulate the female and subdue her aggressive tendency.

In all these types of courtship insemination is effected by the pedipalpi, either simultaneously or in succession. This may be described as mating or direct internal fertilization, one of the many adaptations to terrestrial life of animals whose aquatic ancestors were accustomed to shed their sperm into the water.

With the approach to the second question, why is there any court-ship, Platnick sees in it an important factor in evolution. This follows Mayr's opinion that a change of behaviour accompanies any entry into a new environmental niche, and that other adaptations "particularly the structural ones", are acquired subsequently. Thus it is suggested that courtship methods may be valuable in arranging the genera of a family; for example they may produce a better method of dividing the Salticidae than a system which depends on the number of cheliceral teeth.

If, however, the courtship of spiders marks the crown of the evolu-tionary tree, the lower branches should be considered in an attempt to present the phenomena in a logical order.

The scorpions set the pattern. In their courtship the male first grasps the pedipalpi of the female in his own, and then, tails raised, there follows the well known "promenade à deux". This, interspersed with the actions described as "kissing" and "juddering", last until the pair have reached a place where the ground is smooth. Here the male deposits a loaded spermatophore. The female is then moved so that the drop of spermatozoa is absorbed by her body and fertilization is internal. The essential steps are therefore:

   (i)   the seizing of the female;
  (ii)   the long, preliminary dancing;
 (iii)   the depositing of the spermatophore;
 (iv)   the entry of the sperm-packet.

Reference to the skeleton classification printed among the prelimin-ary pages of this book will show that the scorpions are followed by the Palpigradi and then by the orders formerly included in the "Pedipalpi". Of the courtship of the Palpigradi little or nothing is known, but the Uropygi, which tend to "repeat the scorpion pattern" (see Chapter 15), adopt a courtship that is very similar to that just described. There is the same grasping of the female by the male, the same prolonged dancing which leads to the depositing of a spermatophore.

The Schizomida's courtship closely resembles this, but the next step, which takes us to the Amblypygi, shows a difference. The male does not take hold of the female; instead he touches her long first legs with his own; he even turns his back on her before placing a spermatophore on the ground and leading her to it.

The order most closely related to the Amblypygi is that of the spiders, which have already been fairly fully described. They continue the pro-cess begun by the Amblypygi in that there may be less actual contact between the sexes, and the use of the spermatophore has been abandoned.

In the classification just mentioned the Araneae are followed by five orders that are now extinct and can therefore supply no data by which an interpretation of arachnid courtship might be strengthened. Hence the next orders to be considered are the Opiliones and the Ricinulei.

The former, the harvestmen, are most interesting. An order with an impressive list of adaptations, they are clearly one of the oldest members of the class, and they have had time to produce a complete change in the reproductive behaviour. There is no courtship, no appeasement, no dancing, no spermatophore, no delay. Mating normally follows immediately on the meeting of the sexes, when the male makes use of an intromittent organ, a true penis that is found only in the related Acari.

The Ricinulei are less well known. Pollock (1967) has stated that the male grasps the female, whether or not any courtship has taken place is still uncertain, but the essential fact is that sperm are introduced into the female's body by the third leg of the male and there is no spermatophore.

The position that is thus developing presents first, close contact between the individuals and the use of the spermatophore; then the spermatophore with less physical contact; and lastly the use of a limb or male organ to transfer the spermatozoa.

The fourth sub-class contains but two orders, the Pseudoscorpiones and the Solifugae. One could wish for more orders at this stage of evolution, which may at some distant time have seen some of the missing combinations of characteristics deplored in Chapter 12. However, the data available seem closely to agree with the outline suggested above.

Among false scorpions there are genera in which the traditional clutching of the female occurs, there is a rhythmic walking to and fro, the appearance of a spermatophore, and the impaling of the female upon it. Again, there are genera in which there is lack of close association, and the male abandons the spermatophore, which is later found by the female and all romance has vanished from the insemination.

When, finally, we move to the Solifugae, we find that again the spermatophore has been abandoned, and the packet of spermatozoa is placed in the female orifice by the chelicerae of the male.

Two questions now arise. What is the significance of the dropping of the spermatophore, and, is the courtship of the Arachnida related to the courtship behaviour of other invertebrates?

Courtship is of general occurrence throughout the animal kingdom, and has been fully expounded by Bostock (1969), whose conclusions may be summarized as follows.

(i) It brings male and female together at the right season of the year.

(ii) It ensures that mating takes place in an environment that will suit the needs of the offspring.

(iii) It so synchronized physiological processes and mental moods that both sexes may be excited and neither be aggressive.

(iv) It precludes possibilities of cross-fertilization.

(v) It keeps rival males at a distance.

All this is perfectly true, and conclusions (iii) and (iv) are particularly applicable to the Arachnida. By conclusion (iii) there is implied a vitally important factor, the matter of time.

The result to which these courtship activities usually lead is the amphimixis or union of ova and spermatozoa. In many marine organisms this is made probable by the liberation of the sperm into the water at the moment that the ova are shed, and the courtship helps to ensure that these acts occur simultaneously, and when male and female are close together. In the Arthropoda sperm is usually transferred to the female in the form of a spermatophore, and in this way the semen is not diluted by the surrounding medium. This device was not an arachnid invention. It is to be found among some of the Annelida; it is, as it were, hinted at by the Onychophora; is absent from the Pycnogonida and reappears among the Athropoda. It is therefore a primitive biological feature.

In its simplest form the arachnid spermatophore is a mucilaginous mass of gametes, transferred to the female by the male using his chelicerae, pedipalpi or legs. This type of sperm-transfer is found in the Solifugae and in some Acari. Indirect sperm-transfer by means of complex spermatophores deposited on the ground occurs among scorpions, pseudoscorpions, Amblypgi, Uropygi, Schizomida and some Acari. In these cases the main function of the spermatophore is to prevent desiccation of the semen, and in these orders mating is preceded by a nuptial dance. In this dance the male mesmerizes the females into a position in which she can gather the spermatozoa into her genital orifice. In spiders* and Ricinulei the spermatophore has been lost, and the sperms are transferred by the palpal organs or the modified third leg of the male respectively.

The orders of Arachnida in which a spermatophore is generally used tend to show another common feature, namely an inclination towards a cryptozoic mode of living, which is in itself a primitive stage of evolution. In clear contrast to these are the orders in which other methods are

---

* Cooke (1969) discussing the phylogeny of the male pedipalp, has figured its original condition, with a functional claw "which may have been used to transfer the spermatophore to the female".

found, the Araneae, Opiliones, Ricinulei and Solifugae. These include the most advanced or specialized orders, that is to say, those that have the longest list of characteristics peculiar to themselves. This supports the idea that the spermatophore, valuable and efficient as it is, has been abandoned by these orders; and these are also orders in which a more open or phanerozoic life above the ground layer, in the forests or even in the deserts is the rule.

Entry upon a new environment can seldom succeed unless some new adaptive organ or other feature appears at the critical time, and this may well be a part of the reason for the disappearance of the spermatophore. How the changes took place are matters which, like many others, remain part of the unknown factors of evolution.

38

Arachnophobia

Many of the Arachnida inspire fear; scorpions are commonly avoided as if they were more dangerous than snakes, Solifugae are shunned, and countless people are terrified of spiders. These facts are interesting because there is no obvious justification for so disproportionate, so widespread, so illogical a horror, which may well have been an obstacle to the progress of serious arachnology. What is the reason for and the origin of this fear?

What is fear? Fear is of two kinds. Subjective fear, or, in Spencer's words, "the revival on a given stimulus of past experience of pain", is the consciousness of the changes in blood pressure and muscle tone consequent upon the activity of the adrenal glands, and is therefore rightly described as "unfled flight". Objective fear is the actual secretion of adrenalin, and is an unconditioned reflex normally elicited by an unexpected noise, a sudden shock, or a loss of support, threatening stability of posture.

The tonic reflexes which maintain equilibrium respond to tactile impulses from the soles of the feet, to impulses of pressure from the semicircular canals of the ear, and to visual impulses from the retinas of the eyes. Here we are concerned only with the last of these. From them there arises a tendency, well developed in the lower animals, to keep the images of moving bodies on the same place in each retina. This results in an orientation and movement of their bodies known as rheotropism or rheotaxis.

It is clear that if the body is moved or the eye is turned so as to keep the image of a moving object more or less stationary on the retina, the images of the background and surroundings must, at the same time, be passing across the retina. This is exactly what happens during an unexpected fall and the primitive response is a secretion of adrenalin to fit the body to meet the sudden change of circumstances. This is fear. Just as, according to the James-Lange theory of the emotions we "feel sorry because we cry", so do we feel frightened because of the physical changes in blood pressure and muscle tension.

The sudden running of a spider across the floor, of a scorpion across the tent-cloth or of a solifuge across the desert sand attracts the eye, and as the attempt is automatically made to "keep the eye on" the moving animal, images of the background pass across the retina. The reflex response follows, and, since all motion is relative, the conscious mind has no difficulty in projecting the origin of fear to the escaping creature.

This explanation of the phenomenon of arachnophobia has received criticism during the past three decades, but little of the criticism has been constructive, offering an alternative hypothesis.

During the same period, however, phobias of all kinds, and there are many, have been officially included in the International Classification of Diseases, and considerable advances have been made in understanding and treating them.

Animal phobias, which include a fear of birds, moths, butterflies, wasps, mice, snakes, cats and dogs, have usually been found to have appeared first in childhood. When 18,000 children answered a BBC question, "Which animal do you dislike most?", the snake came first, with 27% of the replies, the spider second with 9·5% and well ahead of the third animal, the lion, which scored 4·5%.

The childish dislike may persist into adult life, when it is sometimes overcome by unprejudiced reason showing the patient that it is simply silly; at other times it is allowed to develop into a serious phobia. Statistics reveal that most of the sufferers are women.

During a period of 12 years in which I was concerned with countless women students, I made a practice of asking them whether they were afraid of spiders, and if so, why? Their replies fell into three groups, each with its qualification: (i) fear of black spiders, not of the gaily-coloured ones; (ii) fear of large spiders, not of the little ones; and (iii) fear of long-legged spiders, not of those with short legs. In every case the answer provoked a second question: just how dark, or how large, or how long in the leg must a spider be before it passes the threshold and becomes frightening? Is there a critical value for each characteristic?

There was never a precise answer to any of these questions, but there was unmistakably a general agreement that colour, size and legginess are all contributory causes, and that most certainly movement cannot be discounted: in fact it becomes clear that the phobia is complex and no one explanation will cover all cases.

One particularly significant confession by a very clever woman traced her fear back to a toy, a papier maché spider with legs of spirally coiled wire. When the spider was hung up, all its legs trembled in a characteristically purposeless manner, inexplicable and terrifying. To this may be added a case, reported to me, of a vertebrate zoologist who almost always referred to the Arthropoda not by name but as "those dam' things with legs".

Extreme cases of arachnophobia present a much more serious aspect. One self-confessed sufferer has written, "I couldn't even write the word spider. I daren't put my handbag on the floor in case a spider crawled over it. . . . I couldn't go into a room until someone had made sure there were no spiders in it". This last limitation is one of the commonest to be reported.

The argument for both speed and legginess as causes of arachnophobia may be taken as established, if not as explained. There is a further aspect of the problem which should not be omitted. Scorpions are among the slowest of the arachnids, Solifugae are certainly the most rapid, and yet among native races they are dreaded with equal intensities. The most probable reason for this is inaccurate information offered and accepted in childhood. Fragmentary knowledge, derived from experience, that certain animals are poisonous, ferocious or otherwise formidable, has been passed on and uncritically transferred to harmless members of the group, with the general result that has evoked the whole of this section. It may surely be paralleled by the number of one's otherwise normal friends who are scared when a wasp flies into the room, only because one day, long ago, they had been tersely informed that "wasps sting". It would be most interesting to know the proportion of arachnophobiacs who could trace their fears back to a casual warning in the nursery.

After the realization that all sufferers from phobias can be accurately described as genuinely ill, both physicians and psychiatrists have sought for alleviation or cure. Some success has followed the training of patients in "deep muscular relaxation", which when continued for a number of sessions produced striking improvements in 90% of the cases treated. Others have been given an injection which induced a state of relaxation, during which the patient was encouraged to imagine himself in a series of increasingly frightening situations and to associate these with a complete absence of fear. In a good proportion of cases a lasting cure resulted.

Psychiatrists have tried to explain the origin of a phobia and to remove it by putting their explanations before the sufferer. The simplest cause is a frightening experience in childhood, and this is admissible because many other phobias have been traced to such an origin. It is undeniable that few creatures are more likely than a house spider to appear unexpectedly and give a shock to a child.

An alternative explanation, which is in keeping with modern psychoanalysis, is to relate the phobia to sex. It is here that the contribution of "long hairy legs" is said to be relevant. Hair has a markedly sexual significance, so that fear and repression of sex may find some relief in an expressed fear of a hairy object. A spider is doubly unfortunate for the "hair" is to its legs what the legs are to its body, so that it has but a poor chance of avoiding its fate as a scapegoat for psychopathetic mentalities. It may be significant that many people who fear spiders can look, unmoved and even interested, at harvestmen, whose conspicuously long legs are almost smooth.

On the other hand, recognition has lately been given to Tipulophobia, the exaggerated fear of the familiar daddy-long-legs or crane fly, an insect whose legs are long but not "hairy". The distinction may perhaps be traced to the insect's body, which is a longish cylinder, unlike the rotund body of a harvestman, and so recalls the fear of the snake, or the tail of a rat and other cylindrical objects.

But long legs can never be regarded as negligible in this respect. Everyone has heard of the Cornish prayer

> "From ghoulies and ghosties and long-leggety beasties,
> And things that go bump in the night,
>   Good Lord, deliver us."

There is, finally, little doubt that the phenomenon of arachnophobia, in its true sense, is not a unified one, and that probably each particular case and each sufferer has its own individual history.

39

Arachnida in Amber

Amber or succinite is the hardened hydrocarbon mixture exuded as resin from the bark of coniferous trees, and especially the extinct species *Pinus succinifer*. It is found in Lower Oligocene beds of sand and loam, and is mined near Palmnicken. It is cast up in quantities on the Baltic coasts and in smaller quantities elsewhere.

It is a substance that has aroused interest for many centuries, first for no cause more subtle than the attractiveness of its appearance, which has brought it within the category of semi-precious materials. Secondly, it is so easily electrified by gentle rubbing that its name *elektron* in ancient Greece is the origin of the word "electricity"; and thirdly, as was known to Tacitus, many insects and other small animals were captured by its stickiness when it was freshly exposed and, covered by it, have been preserved for us to examine many millions of years after their death. Arachnida are included among these.

The first monographic treatment of amber spiders was a paper by Koch and Behrendt, edited by Menge and published in 1854. Animals that are inaccessible in the middle of a block of translucent resin are not easily examined, and the consequence was an inevitable imperfection in Koch's descriptions. Moreover, the state of the systematics of spiders a century ago was such that the method of describing any new species was less precise than is now found to be necessary, with the consequence that even with Menge's careful editing and annotating, most of Koch's species are now indeterminable. Since the original specimens appear all to have been lost there is now no chance of a revision.

Other descriptions of amber-encased spiders in the following decades were sporadic and fragmentary and added little to our knowledge of the Oligocene fauna: there was in fact no significant advance until Petrunkevitch began his studies of amber spiders, now incorporated in four papers.

The first of these, "A Study of Amber Spiders", was published in

1942. An examination of 144 specimens yielded 78 named species, 69 of which were new, in 27 families. A most important part of the paper was Petrunkevitch's description of his techniques for improving the visibility of the specimen. He tells us how removal of the surface of a piece of amber often displays much more transparent material below the darkened edges, how the effect of bubbles can be overcome by drilling down to them and filling the cavity with liquid; and how the improved object can be mounted and examined. His ability as a photographer is nowhere better shown than in this work on amber, and his papers include a large number of admirable photographs supplemented by clear explanatory drawings. This paper also contained a discussion of the evolutionary changes in spiders' bodies.

The publications of 1942 and 1950 were applications of these methods and principles to specimens of amber in the possession of the American Museum of Natural History and the Museum of Comparative Zoology at Harvard University. They added 38 new species to the list of named amber spiders.

The paper of 1958 was on a larger scale. Entitled "Amber Spiders in European Collections", it dealt with 230 specimens sent on loan for the purposes of study by the museums and universities of almost all the interested European states. The result was a full description of them all and an addition of 47 new species. The paper ended with a general summary of the work of the 17 years in the form of a list of all the known species of amber spiders, complete with synonymy. This gave a total of 274 species, placed in 27 families.

Six of these families were new, and represented lines of evolution now extinct. They were:

Adjutoridae	3 species
Arthrodictynidae	1 species
Ephalmatoridae	2 species
Inceptoridae	2 species
Insecutoridae	3 species
Spatiatoridae	3 species.

Harvestmen have also been included in amber, and an account of them was published by Roewer in 1939. The pioneer work of Koch and Behrendt, already mentioned, contained descriptions of four species of the sub-order Palpatores:

Nemastoma denticulatum
Caddo dentipalpis
Dicranopalpus ramiger
Opilio ovalis.

Menge in 1854 added two more:

Cheiromachus corriaceus
Leiobunum longipes

and Roewer brought the total to seven with

Subacon bachofeni.

Seven is not a large proportion of the 1,500 or so Palpatores alive today, but it is greater than the fossil representation of the sub-order Laniatores, of which but a single species, *Gongyleptes nemastomoides*, was described by Koch and Behrendt.

False scorpions have left a rather better record. There are more than a dozen families of living Pseudoscorpiones, and nine of these are represented by 18 species in Baltic amber. There is also a description of a Miocene species from Burma.

It may be noted that in both Araneae and Pseudoscorpiones the number of fossil species known is about 1% of the total: in Opiliones it is less than one-fifth of this.

VI. EPILEGOMENA

Bibliography

I. General Works

ARACHNIDA

Berland, L., "Les Arachnides". Paris, 1932.
Cloudsley-Thompson, J. L., "Spiders, Scorpions, Centipedes and Mites". London, 1968.
Comstock, J. H., "The Spider Book". New York, 1948.
Grassé, P.-P. (ed.), "Traité de Zoologie", Tome VI. Paris, 1949.
Hansen, H. J. and Sørensen, W., "On Two Orders of Arachnida". Cambridge, 1904.
Petrunkevitch, A., in Moore, R. C. ed. "A Treatise on Invertebrate Palaeontology". Kansas, 1955.
Savory, T. H., "Spiders, Men and Scorpions". London, 1961.
Simon, E., "Les Arachnides de France". Paris, 1874–1937.
Snow, K. R., "The Arachnids: An Introduction". London, 1970.
Vachon, M., "Chelicerates" in Encyclopédie de la Pléiade, Zoologie, Tome II. Paris, 1963.
Warburton, C., "Arachnida Embolobranchiata" in Cambridge Natural History. London, 1909.

ARANEAE

Baerg, W. J., "The Tarantula". Kansas, 1958.
Berland, L., "Les Araignées". Paris, 1938.
Bristowe, W. S., "The Comity of Spiders". London, 1939–41.
Bristowe, W. S., "The World of Spiders". London, 1958.
Child, J., "Australian Spiders". Melbourne, 1968.
Forster, R. R. et al., "Spiders of New Zealand". Dunedin, 1967– .
Forster, R. R. and Forster, L. M., "New Zealand Spiders". Auckland, 1973.
Gertsch, W. J., "American Spiders". New York, 1949.
Kaston, B. J., Spiders of Connecticut. Bull. Conn. geol. nat. Hist. Surv. No. 70:1–874, 1948.
Locket, G. H., Millidge, A. F. and Merrett, P., "British Spiders". London, 1951–74.
McKeown, K. C., "Australian Spiders". Sydney, 1952.
Nielsen, E., "The Biology of Spiders". Copenhagen, 1932.
Savory, T. H., "The Spider's Web". London, 1952.
Savory, T. H., "The Biology of Spiders". London, 1920.
Simon, E., "Histoire Naturelle des Araignées". Paris, 1892–1903.
Thorp, R. W. and Woodson, W. D., "Black Widow". Chapel Hill, N. Carolina, 1945.
Tilquin, A., "La Toile Géométrique des Araignées". Paris, 1942.

Vellard, J., "Le Venin des Araignées". Paris, 1936.
Witt, P. N. *et al.*, "A Spider's Web". New York, 1938.
Yaginuma, T., "Spiders of Japan in Colour". Osaka, 1968.
Yates, J. H., "Spiders of Southern Africa". Cape Town, 1968.

ACARI

Arthur, D. R., "Ticks and Disease". London, 1962.
Arthur, D. R., "British Ticks". London, 1963.
Baker, E. W. and Wharton, G. W., "An Introduction to Acarology". New York, 1952.
Evans, G. O., Sheals, J. G. and Macfarlane, D., "The Terrestrial Acari of the British Isles". Vol. 1. Brit. Mus. (N.H.), London, 1961.
Evans, G. O. and Till, W. M., Studies on the British Dermanyssidae. *Bull. Brit. Mus. nat. Hist.* Zool. **13,** 249–294. 1965.
Hughes, A. M., "The Mites of Stored Food". London, 1961.
Krantz, G. W., "A Manual of Acarology". Corvallis, Oregon, 1970.

OTHER ORDERS

Beier, M. Pseudoscorpionidea. *Tierreich* Lief. 57, 58. Berlin and Leipzig, 1932.
Berland, L., "Les Scorpions". Paris, 1948.
Bonnet, P. "Bibliographia Araneorum", 3 vols. Toulouse, 1945–61.
Chamberlin, J. C., "The Arachnid Order Chelonethida". Stanford 1931.
Kolosváry, G., "Die Weberknechte Ungarns". Szeged, 1929.
Roewer, C. F., "Die Weberknechte der Erde". Jena, 1923.
Roewer, C. F. Solifugae. Bronn's "Klassen und Ordnungen des Tierreichs", **5,** IV, Buch 4. Leipzig, 1933.
Sankey, J. H. P. and Savory, T. H., British Harvestmen. *Synopses Br. Fauna*, N.S., No. 4: 1–76. London, 1974.
Vachon, M., "Études sur les Scorpions". Alger, 1952.
Weygoldt, P., "The Biology of Pseudoscorpions". Cambridge, Mass., 1969.

Acta Arachnologica. Osaka. *passim.*
Bulletin of the British Arachnological Society. passim.
Journal of Arachnology. Texas. *passim.*

II. References in the Text

Alexander, A. J. (1957). The courtship and mating in the scorpion *Opisthophthalmus latimanus. Proc. zool. Soc. Lond.* **128,** 529–44.
Alexander, A. J. and Ewer, D. W. (1957). On the origin of mating behaviour in spiders. *Amer. Nat.* **91,** 311–27.
Alexander, A. J. (1958). On the stridulation of scorpions. *Behaviour* **xii,** 339–52.
Anderson, J. F. (1974). Responses to starvation in the spiders *Lycosa lenta* and *Filistata hibernalis. Ecology* **55,** 576–85.
Anderson, S. O. (1971). Aminoacid composition of spiders' silk. *Comp. Biol. Physiol.* **35,** 705–11.
André, M. (1932). La sécrétion de la soie chez les acariens. "Livre du centenaire." *Soc. ent. Fr.,* 457–73.

Apstein, G. (1889). Bau und Function der Spinndrüsen der Araneidea. *Arch. Naturgesch.* **55**, 29–74.

Barrows, W. M. (1915). The reactions of an orb-weaving spider to rhythmic vibrations of its web. *Biol. Bull.* **29**, 316–32.

Barth, F. G. (1967). Ein einzelnes Spaltsinnesorgan auf dem Spinnentarsus. *Z. vergl. Physiol.* **55**, 407–49.

Barth, F. G. and Seyfarth, E. A. (1972). Compound slit sense organs on the spider's leg. *J. Comp. Physiol.* **78**, 176–91.

Barth, F. G. and Wadepuhl, M. (1975). Slit sense organs in the scorpion leg. *J. Morphol.* **145**, 209–28.

Berland, L. (1914). Un palpigrade nouveaux trouvé dans les serres du Muséum national d'Histoire naturelle. *Bull. Soc. ent. Fr.* No. 12, 375–377.

Bishop, S. C. (1945). Our lady's threads. *Trans. Conn. Acad. Arts Sci.* **36**, 91–7.

Blum, M. S. and Edgar, A. L. (1971). 4-Methyl-3-heptanone in opilionid exocrine secretions. *Insect Biochem.* **1**, 181–188.

Blumenthal, R. (1935). Untersuchungen über das "Tarsal-organ" der Spinnen. *Z. Morphol. Ökol. Tiere* **29**, 667–719.

Bonnet, P. (1930). La mue, l'autotomie et a régénération chez les araignées. *Bull. Soc. Hist. nat. Toulouse* **59**, 237–700.

Bonnet, P. (1935). Araignées mâles à palpe unique. *Bull. Soc. Hist. nat. Toulouse* **68**, 411–14.

Börner, C. (1921). Die Gliedmassen der Arthropoden. *Handb. Morph. wirb. Tiere* **iv**, 649–94.

Bostock, C. (1969). "Courtship in Animals." London.

Braendegaarde, J. (1946). The spiders of East Greenland. *Medd. Grønland* **121**, 15, 1–128.

Brauer, A. (1917). Ueber Doppelbildungen des Skorpions (*Euscorpius carpathicus* L.), *S.B. Akad. Wiss. Wien.* 208–21.

Briggs, T. S. (1969). A new Holarctic family of Laniatorid Phalangida. *Pan-Pacific Ent.* **45**, 35–50.

Bristowe, W. S. (1930). The distribution and dispersal of spiders. *Proc. zool. Soc. Lond.* 633–57.

Bristowe, W. S. (1933). The Liphistiid spiders. *Proc. zool. Soc. Lond.* 1015–57.

Bristowe, W. S. (1949). The distribution of harvestmen in Great Britain and Ireland. *J. Anim. Ecol.* **18**, 100–14.

Bristowe, W. S. and Locket, G. H. (1926). The courtship of British lycosid spiders. *Proc. zool. Soc. Lond.* 317–47.

Buxton, B. H. (1913). Coxal glands of the arachnids. *Zool. Jahrb.* **14**, 231–82.

Cambell, F. M. (1882). On a probable case of parthenogenesis in the House Spider. *J. Linn. Soc.* **16**, 536–9.

Chamberlin, R. V. and Ivie, W. (1933). Spiders of the Raft River Mountains. *Bull. Univ. Utah* **23**, 1–53.

Chamberlin, R. V. and Ivie, W. (1936). Arachnida of the orders Pedipalpida, Scorpionida and Ricinulida. *Publ. Carneg. Inst.* **491**, 101–7.

Cloudsley-Thompson, J. L. (1973). Entrainment of the "Circadian clock" in *Buthotus minax. J. Interdiscipl. Cycle Res.* **4**, 119–23.

Coad, B. R. (1931). "Aerial Plankton." Year book of Agriculture, N. York.

Condé, B. (1965). Présence de Palpigrades dans le milieu interstitial littoral. *C. r. hebd. Séanc. Acad. Sci. Paris* **261**, 1898–1900.

Cooke, J. A. L. (1970). Spider genitalia and phylogeny. *Bull. Mus. Hist. nat., Paris* **41** Suppl., 142–6.

Cooke, J. A. L. (1972a). The urticating hairs of Theraphosid spiders. *Amer. Mus. Novit.* **2498**, 1–34.

Cooke, J. A. L. (1972b). Stinging hairs: a tarantula's defense. *Fauna* **4**, 4–8.

Cooke, J. A. L. and Shadab, M. U. (1973). Whipscorpions from Africa. *Amer. Mus. Novit.* **2526**, 1–11.

Crosby, C. R. (1934). An interesting two-eyed spider from Brazil. *Bull. Brooklyn ent. Soc.* **29**, 19–23.

Curtis, D. J. (1970). Comparative aspects of the fine structure of the eyes of Phalangida. *J. Zool., Lond.* **160**, 231–65.

Dahl, F. (1911). Die Hörhaare und das System der Spinnentiere. *Zool. Anz.* **37**, 522–32.

Damin, N. (1894). On parthenogenesis in spiders. *Ann. Mag. nat. Hist.* **6**(14), 26–9.

Dondale, C. D. and Hegdekar, B. M. (1973). The contact sex pheromone of *Pardosa lapidicina. Can. J. Zool.* **51**, 400–1.

Dondale, C. D. and Legendre, R. (1971). Mise en évidence de phénomènes de diapause hivernale chez l'araignée paléarctique *Pisaura mirabilis. C. R. Acad. Sci. Paris.* **270**, 2483–5.

Dubinin, W. B. (1957). New system of the superclass Chelicerata. *Bull. Soc. mosc. exp. Nat.* **62**, 25–33.

Duffey, E. (1968). An ecological analysis of the spider fauna of sand dunes. *J. Anim. Ecol.* **37**, 461–74.

Duffey, E. (1972). Ecological survey and the arachnologist. *Bull. Br. arachnol. Soc.* **2**, 69–82.

Edgar, A. L. (1963). Proprioception in the legs of phalangids. *Biol. Bull. mar. biol. Lab. Woods Hole* **124**, 262–7.

Edgar, A. L. and Yuan, H. A. (1968). Daily locomotory activity in *Phalangium opilio. Bios.* **39**, 167–76.

Eisner, T. J., Meinwald, J., Munro, A. and Ghent, R. (1961). Defence mechanisms of arthropods. *J. Insect. Physiol.* **6**, 272–98.

Emerton, J. H. (1919). The flights of spiders in the autumn of 1918. *Ent. News* **30**, 165–8.

Ewing, H. E. (1929). A synopsis of the American arachnids of the primitive order Ricinulei. *Ann. ent. Soc. Amer.* **22**, 583–600.

Fabre, J. H. (1913). "The Life of the Spider." London.

Firstman, B. (1973). The relationship of the chelicerate arterial system to the evolution of the endosternite. *J. Arachn.* **1**, 1–54.

Fischer, E. (1907). Ueber Spinnenseide. *Z. physiol. Chem.* **53**, 126–39.

Forster, R. R. (1949). The sub-order Cyphophthalmi in New Zealand. *Dom. Mus. Ent. Rec.* **1**, 79–119 and 179–211.

Fowler, D. J. and Goodnight, C. J. (1966a). The cyclic production of 5-hydroxytryptamine in the opilionid. *Amer. Zool.* **6**, 187–93.

Fowler, D. J. and Goodnight, C. J. (1966b). Neurosecretory cells; daily rhythmicity in *Leiobunum longipes. Science, N.Y.* **152**, 1078–80.

Gabbutt, P. D. (1972). Differences in the disposition of trichobothria in the Chernetidae. *J. Zool., Lond.* **167**, 1–13.

Gabbutt, P. D. and Vachon, M. (1967). The external morphology and life history of the pseudoscorpion *Roncus lubricus. J. Zool., Lond.* **153**, 475–98.

Gabbutt, P. D. and Vachon, M. (1968). The external morphology and life history of the pseudoscorpion *Microcreagris cambridgei. J. Zool., Lond.* **154**, 421–41.

Gerhardt, U. (1921). Vergleichende Studien über die Morphologie des männlichen Tasters und die Biologie der Kopulation der Spinnen. *Arch. Naturgesch.* **87**, 78–247.

Gerhardt, U. (1924). Neue Studien zur Sexualboilogie der Spinnen. *Z. Morph. Ökol. Tiere* **1,** 507–38.

Gilbert, O. (1951). Observations on the feeding of some British false scorpions. *Proc. zool. Soc. Lond.* **121,** 547–55.

Grassi, B. (1886). I progenitori dei Miriapodi e degli Insetti. Memoria v, Intorno ad un nuovo Aracnide artrogastro (*Koenenia mirabilis*). *Boll. Soc. ent. ital.* **18,** 153–72.

Guenthal, J. (1944). Du développement postembryonnaire de *Phalangium opilio*. *Bull. Soc. zool. Fr.* **68,** 98–100.

Hegdeker, B. M. and Dondale, C. D. (1969). A contact sex pheromone in lycosid spiders. *Canad. J. Zoo.* **47,** 1–4.

Henriksen, K. L. and Lundbeck, W. (1918). Landarthropoder (Insecta et Arachnida). *Meddr. Grønand.* **22,** 481–823.

Hickman, V. V. (1931). A new family of spiders. *Proc. Zool. Soc., Lond.* **1931,** 1321–8.

Hickman, V. V. (1967). "Spiders of Tasmania." Tasman Museum, Hobart.

Hingston, R. W. G. (1928). "Problems of Instinct and Intelligence." London.

Holzapfel, M. (1933). Die nicht-optische Orientierung der Trichterspinne *Agelena labyrinthica* (CL). *Z. vergl. Physiol.* **20,** 55–116.

Homann, H. (1928). Beiträge zur Physiologie der Spinnenaugen. *Z. vergl. Physiol.* **7,** 201–67.

Homann, H. (1971). The eyes of Araneae. *J. morphol. Tiere* **69,** 201–72.

Hunt, S. (1970). Amino acid composition of silk from the pseudoscorpion *Neobisium maritimum*. *Comp. Biochem. Physiol.* **34,** 773–6.

Hunt, S. and Legg, G. (1971). Protein component of the spermatophore of the pseudoscorpion *Chthonius ischnocheles*. *Comp. Biochem. Physiol.* **35,** 457–9.

Jackson, R. R. (1973). Nomenclature for orb web thread connections. *Bull. Br. arachnol. Soc.* **2,** 125–6.

Juberthie, C. (1961). Structure des glandes odorantes et modalités d'utilisation de leur sécrétion chez deux opilions Cyphophthalmes. *Bull. Soc. zool. Fr.* **86,** 106–16.

Juberthie, C. (1963). Étude des Ópilions Cyphophthalmes. *Bull. Mus. Hist. nat., Paris* **34,** 267–75.

Juberthie, C. (1964). Recherches sur la biologie des Opilions. Thesis: Faculté des Sciences, Université de Toulouse. (also *Ann. Spéléol.* **19,** 1–238).

Juberthie, C. (1967a). Siro rubens. *Revue Ecol. Biol. Sol* **4,** 155–71.

Juberthie, C. (1967b). Caractères sexuelles secondaires des Opilions: les glandes anales de *Siro rubens*. *Revue Ecol. Biol. Sol.* **4,** 489–96.

Juberthie, C. (1968). Une nouvelle espèce de Cyphophthalmes de Grèce. *Revue. Ecol. Biol. Sol.* **3,** 549–59.

Junqua, C. and Vachon, M. (1968). Les Arachnides venimeux et leurs venins. *Mém. Acad. r. Sci. outre-mer Cl. Sci. nat. méd. 8°,* N.S. **17** (5), 1–136.

Kästner, A. (1931). Die Hüfte und ihre Umformung zu Mundwerkzeugen bei den Arachniden. *Z. Morph. Ökol. Tiere* **22,** 721–58.

Kaston, B. J. (1974). Remarks on the names of families and higher taxa in spiders. *J. Arachnol.* **2,** 47–52.

Klee, G. E. and Butcher, J. W. (1968). Laboratory rearing of *Phalangium opilio*. *Mich. Ent.* **1,** 275–8.

Koch, C. L. and Berendt, G. C. (1854). In "Die im Bernstein befindlichen Organischen Reste der Vorwelt" (Berendt, G. C., ed.). **1,** Berlin.

Kraepelin, K. (1899). Scorpiones und Pedipalpi. *Tierreich* **8,** 1–265.

Kraepelin, K. (1901). Palpigradi und Solifugae. *Tierreich* **12,** 1–159.

Kühn, A. (1919). "Die Orientierung der Tiere im Raum." Jena.

Lamoral, B. H. (1968). On the nest and web structure of *Latrodectus* in South Africa. *Ann. Natal Mus.* **20**, 1–14.

Lamoral, B. H. (1973). Anatomy of the tarsal organ of *Palystes natalius*. *Ann. Natal Mus.* **21**, 609–48.

Lamoral, B. H. (1975). The structure and possible function of the flagellum in four species of male Solifuges. *Proc. 6th Int. Arach. Cong., Amsterdam* 136–40.

Lankester, E. R. (1904). The structure and classification of the Arachnida. *Q. Jl. microsc. Sci.* **48**, 165–269.

Lawrence, R. F. (1928). Contributions to a knowledge of the fauna of South-West Africa. *Ann. S. Afr. Mus.* **25**, 217–312.

Lawrence, R. F. (1931). The Harvest-spiders of South Africa. *Ann. S. Afr. Mus.* **29**, 341–508.

Lawrence, R. F. (1947). Some observations on the eggs and newly-hatched embryos of *Solpuga hostilis*. *Proc. zool. Soc. Lond.* **117**, 429–34.

Lawrence, R. F. (1953). "The Biology of the Cryptic Fauna of Forests, with Special Reference to the Indigenous Forests of South Africa." Cape Town.

Lawrence, R. F. (1954). Fluorescence in arthropoda. *J. ent. Soc. S. Afr.* **17**, 167–70.

Legendre, R. (1956). Les éléments neurosécréteurs de la masse nerveuse et leur cycle d'activité chez les aranéides. *Comptes rendus Acad. Sci. Paris* **242**, 2225–6.

Legendre, R. (1963). L'audition et l'émission de sons chez les aranéides. *Ann. Biol.* **2**, 371–90.

Legendre, R. (1965). Morphologie et développment des Chélicérates. *Fortschr. Zool.* **17**, 239–71.

Legendre, R. (1967). Morphologie und Entwicklungsgeschichte der Cheliceraten. *Fortschr. Zool.* **18**, 207–22.

Legendre, R. (1968). Morphologie et développement des Chélicérates. *Fortschr. Zool.* **19**, 1–50.

Legg, G. (1973a). Spermatophore formation in the pseudoscorpion *Chthonius ischnocheles*. *J. Zool., Lond.* **170**, 367–94.

Legg, G. (1973b). The structure of encysted sperm of some British Pseudoscorpiones. *J. Zool., Lond.* **170**, 429–40.

Legg, G. (1975). The possible significance of spermathecae in Pseudoscorpionas. *Bull. Br. arachnol. Soc.* **3**, 91–5.

Lehtinen, P. T. (1967). Classification of the Cribellate spiders. *Annls zool. fenn.* **4**, 199–468.

Levi, H. (1967). Adaptations of respiratory systems of spiders. *Evolution* **21**, 571–83.

Loeb, J. (1916). "The Organism as a Whole." New York.

Loeb, J. (1918). "Forced Movements, Tropisms and Animal Conduct." Philadelphia and London.

Lopez, A. (1972). Morphologie des glandes epigastriques dans les Dysderidae et Clubionidae. *Bull. Soc. zool. Fr.* **97**, 113–19.

Lopez, A. (1973). La structure fine des glandes epigastriques et les glandes de soie chez *Pholcus phalangioides*. *C. R. Acad. Sci., Paris* **276**, 2681–4.

Machado, A. de B. (1964). Sur l'existence de la parthénogenèse dans quelques espèces d'araignées ochyroceratides. *C. r. hebd. Séanc. Acad. Sci. Paris* **258**, 5059–9.

Manton, S. (1973). Arthropod phylogeny—a modern synthesis. *J. Zool., Lond.* **171**, 111–30.

Maretié, Z. (1975). European araneism, *Bull. Br. arachnol. Soc.* **3**, 126–30.

Marples, B. J. (1967). The spinnerets and epiandrous glands of spiders. *J. Linn. Soc. Zool.* **46**, 209–22.

Marples, B. J. and Marples, M. J. (1971). Notes on the behaviour of spiders in the genus *Zygiella*. *Bull. Br. Arach. Soc.* **2**, 16–17.

Mathew, A. P. (1957). Mating of scorpions. *J. Bombay nat. Hist. Soc.* **54**, 853–7.

Mathew, A. P. (1961). Embryonic nutrition in *Lychas tricarinatus*. *J. zool. Soc. India* **12**, 220–8.

Menge, A. (1854). See Koch and Berendt.

Millot, J. (1926). Contributions a l'histophysiologie des Aranéides. *Bull. biol. Fr. Belg.* Suppl. **8**, 1–238.

Millot, J. (1929). Les glandes séricigènes des Pholcides. *Bull. Soc. Zool. Fr.* **54**, 194–206.

Millot, J. (1931a). Les glandes séricigènes des Dysdérides. *Arch. Zool. exp. gén.* **71**, 38–45.

Millot, J. (1931b). Anatomie comparée de l'intestin moyen céphalothoracique chez les araignées vraies. *C. R. Acad. Sci., Paris.* **192**, 375–7.

Millot, J. (1931c). Les diverticules intestinaux du céphalothorax chez les araignées vraies. *Z. Morph. Ökol. Tiere* **21**, 740–64.

Millot, J. (1942–3). Sur l'anatomie de *Koenenia mirabilis*. *Rev. franç. Ent.* **9**, 33–51 and 127–35.

Millot, J. (1945a). L'anatomie interne des Ricinulei. *Annls. Sci. nat. (Zool.)* **7**, 1–29.

Millot, J. (1945b). Les Ricinulei ne sont pas des arachnides primitifs. *Bull. Soc. zool. Fr.* **70**, 106-8.

Mitchell, R. W. (1968). *Typhlochactas* a new genus of eyeless cave scorpion from Mexico. *Ann. Spéléologie* **23**, 753–77.

Mitchell, R. W. (1970). A Mexican cavernicole Ricinuleid. *Ciencia.* **27**, 63–74.

Moorhead, C. (1970). Arachnophobia. Daily Telegraph Magazine **307**.

Morel, G. (1974a). Étude de l'action de bacillus thuringiensis chez le scorpion *Buthus occitanus*. *Entomophaga* **19**, 85–95.

Morel, G. (1974b). Histopathologie du Scorpion *Buthus occitanus*. *Bull. Soc. zool. Fr.* **99**, 479–97.

Muchmore, W. B. (1971). On phoresy in pseudoscorpions. *Bull. Br. arachnol. Soc.* **2**, 38.

Muma, M. H. (1966). The life cycle of *Eremobates durangonus* (Arachnida: Solpugida). *Florida Entomologist* **49**, 233–42.

Osaki, H. (1969). Electron microscope study of the spermatozoon of the liphistiid spider *Heptathela kimurai*. *Acta arachnologica* **22**, 1–12.

Parry, D. A. and Brown, R. H. J. (1959). The jumping mechanism of salticid spiders. *J. exp. Biol.* **36**, 654–64.

Peters, H. M. (1931–33). Die Fanghandlung der Kreuzspinne (*Epeira diademata* Cl.). *Z. vergl. Physiol.* **15**, 693–748 and **19**, 47–67.

Peters, H. M. (1933). Kleine Beiträge zur Biologie der Kreuzspinne (*Epeira diademata* Cl.). *Z. Morph. Ökol. Tiere* **26**, 447–68.

Peters, H. M. (1947). Zur Geometrie des Spinnen-netzes. *Z. Naturf.* **2b**, 227–33.

Peters, P. J. (1970). Orb web construction. *Anim. Behav.* **18**, 478–84.

Petrunkevitch, A. (1913). A monograph of the terrestrial Palaeozoic Arachnida of North America. *Trans. Conn. Acad. Sci.* **18**, 1–137.

Petrunkevitch, A. (1933). An inquiry into the natural classification of spiders. *Trans. Conn. Acad. Sci.* **31**, 299–389.

Petrunkevitch, A. (1942). A study of Amber Spiders. *Trans. Conn. Acad. Sci.* **34**, 119–464.

Petrunkevitch, A. (1945). Palaeozoic Arachnida of Illinois. *Ill. State Mus. Sci. Pap.* **3**, 1–72.

Petrunkevitch, A. (1949). A study of Palaeozoic Arachnida. *Trans. Conn. Acad. Sci.* **37**, 69–315.

Petrunkevitch, A. (1950). Baltic Amber Spiders in the museum of comparative Zoology. *Bull. Mus. comp. Zool.* **103**, 259–337.

Petrunkevitch, A. (1953). Palaeozoic and Mesozoic Arachnida of Europe. *Mem. geol. Soc. Amer.* No. 53, 1–128.

Petrunkevitch, A. (1958). Amber Spiders in European collections. *Trans. Conn. Acad. Sci.* **41**, 97–400.

Phillipson, J. (1959). The seasonal occurrence, life histories and fecundity of harvest-spiders in the neighbourhood of Durham city. *Ent. mon. Mag.* **95**, 134–8.

Pittard, K. and Mitchell, R. W. (1972). Comparative morphology of the life stages of *Cryptocellus pelaezi*. *Grad. Stud. Texas Tech. Univ.* No. 1, 1–77.

Platnick, N. (1971). The evolution of courtship behaviour in spiders. *Bull. Br. Arach. Soc.* **2**, 40–7.

Platnick, N. and Levi, H. W. (1973). On family names of spiders. *Bull. Br. arachnol. Soc.* **2**, 166–7.

Pocock, R. I. (1911). A monograph of the terrestrial Carboniferous Arachnida of Great Britain. *Palaeontogi. Soc. (Monogi.)* 1911, 1–84.

Pocock, R. I. (1929). Arachnida. *In* "Encyclopaedia Britannica", 16th ed.

Pollock, J. (1967). Notes on the biology of Ricinulei. *Jl W. Afr. Sci. Ass.* **12**, 19–22.

Pringle, J. W. S. (1955). The function of the lyriform organs of arachnids. *J. exp. Biol.* **32**, 270–8.

Reed, C. F., Witt, P. N., Scarboro, M. B. and Peakall, D. B. (1970). Experience and the orb web. *Developmental Psychobiology* **3**, 251–65.

Reiskind, J. (1965). Self-burying behaviour in the genus *Sicarius* (Araneae: Sicariidae). *Psyche* **72**, 218–24.

Roewer, C. F. (1939). Opilioniden im Bernstein. *Palaeobiologica* **7**, 1–5.

Roters, M. (1944). Observations on British harvestmen. *J. Quekett microsc. Club*, **11**, 23–5.

Rovner, J. S. (1966). Courtship in spiders without prior sperm induction. *Science, N.Y.* **152**, 543–4.

Rovner, J. S. (1967a). Copulation and sperm induction by normal and palpless male linyphiid spiders. *Science, N.Y.* **157**, 835.

Rovner, J. S. (1967b). Acoustic communication in a lycosid spider. *Anim. Behav.* **15**, 273–81.

Rovner, J. S., Higashi, G. A. and Foelix, R. F. (1973). Maternal behavior in wolf spiders. *Science, N.Y.* **182**, 152–5.

Rovner, J. S. (1973). Copulatory pattern supports generic placement of *Schizocosa avida*. *Psyche* **80**, 245–8.

Rowland, J. M. and Cooke, J. A. L. (1973). Systematics of the Arachnid order Uropygida. *J. Arachnol.* **1**, 55–71.

Sankey, J. H. P. (1950). British harvest-spiders. *Essex Nat.* **28**, 181–91.

Savigny. J. C. (1816). "Mémoires sur les animaux sans vertèbres." Paris.

Savory, T. H. (1952). "The Spider's Web." London.

Savory, T. H. (1974). On the arachnid order Palpigradi. *J. Arachnnol.* **2**, 43–5.

Savory, T. H. (1975). On the problem of Arachnid evolution. *Proc. 6th Arachn. Congress, Amsterdam,* 46–8.

Savory, T. H. (1977). Cyphophthalmi: the case for promotion. *Biol. J. Linn. Soc.*

Sekiguichi, K. (1952). On a new spinning gland found in geometric spiders and its functions. *Annotnes zool. jap.* **25**, 394–9.

Sekiguichi, K. (1957). Redublication in spider eggs produced by centrifugation. *Sci. Rep. Tokyo Kyoiku Daigaku* **8B**, 227–80.

Spoek, G. I. (1963). The Opilionida of the Netherlands. *Zool. Verh.*, *Leiden* **62**, 1–70.

Stipperger, H. (1928). Biologie und Verbreitung der Opilioniden Nordtirols. *Arb. zool. Inst. Univ. Innsbruck* **3**, 19–79 (3–63).

Størmer, L. (1944). On the relationships and phylogeny of fossil and recent Arachnomorpha. *Vid. Akad. Oslo* **1**, 1–158.

Størmer, L. (1963). *Gigantoscorpius willsi*, a new Scorpion from the Lower Carboniferous of Scotland. *Vid. Akad. Oslo* N.S. **8**, 1–171.

Suzuki, S. (1951, 2, 4). Cytological studies in Spiders. *J. Sci. Hiroshima Univ. Zool.* **12**, 67–98; **13**, 1–52; **15**, 23–136.

Szlep, R. (1966a). Evolution of the web-spinning activities: the web spinning in *Titanoeca albamaculata*. *Israel J. Zool.* **15**, 83–8.

Szlep, R. (1966b). The web structure of *Latrodectus variolus*. *Israel J. Zool.* **15**, 89–94.

Thorell, T. T. T. and Lindstrom, G. (1885). A Silurian Scorpion from Gotland. *K. svenska VetenskAdad. Handl.* **21**, 9.

Tuxen, S. L. (1973). The African genus *Ricinoides*. *J. Arachnol.* **1**, 85–106.

Vachon, M. (1932). Recherches sur la biologie des Pseudoscorpionides. *Bull. sci. Bourgogne* **2**, 21–6.

Vachon, M. (1941). *Chthonius tetrachelatus* et ses formes immatures. *Bull. Mus. Hist. nat., Paris* (2) **13**, 442–9, 540–7.

Vachon, M. (1952). "Études sur les scorpions." Alger.

Vachon, M. (1968). "Piqures et Morsures d'Arthropodes." Cahiers Sandoz.

Vachon, M. (1973). Études des caractères utilisés pour classer les familles et les genres de Scorpions. *Bull. Mus. Hist. nat., Paris* **140**, 857–958.

Vachon, M. (1974). Concept et Rôle du Caractère en Classification. *Proc. 6th Int. Arach. Congress*, Amsterdam, 3–6.

Versluys, J. and Demoll, R. (1922). Das Limulus-Problem. *Ergebn. Fortschr. Zool.* **5**, 67–388.

Warburton, C. (1890). The spinning apparatus of geometric spiders. *Q. Jl Micro. Sci.* **31**, 29–39.

Warren, E. (1926). On the habits of *Palystes natalius*. *Ann. Natal Mus.* **5**, 308–49.

Weygoldt, P. (1971). Notes on the life history and reproductive biology of the giant whip-scorpion *Mastigoproctus giganteus* from Florida. *J. Zool., Lond.* **164**, 137–47.

Weygoldt, P. (1972). Geisselskorpione und Geisselspinnen. *Z.Kölner Zoo.* **15**, 95–107.

Weygoldt, P. (1973). Spermatophorenbau und Samenubertragung bei Uropygen. *Z. Morph Okol. Tiere* **71**, 23–51.

Williams, S. C. (1968). Methods of sampling scorpion population: review and evaluation. *Proc. Calif. Acad. Sci.* **36**, 221–30.

Witt, P. N. (1971). Drugs alter web-building of spiders. *Behavioural Science* **16**, 98–113.

Wood, F. D. (1926). Autotomy in Arachnida. *J. Morph.* **42**, 143–95.

Yoshikura, J. (1969). Effects of ultraviolet irradiation on the embryonic development of a liphistiid spider. *Kumamoto J. Sci.*, B, **9**, 57–108.

Yoshikura, M. (1965). Postembryonic development of a whip-scorpion. *Kumamoto J. Sci.*, B, **7**, 21–50.

Yoshikura, M. and Takano, S. (1972). Neurosecretory system of the purse-web spider. *Kumamoto J. Sci.*, B, **11**, 29–36.

Zapfe, H. (1961). La familia Migidae en Chile. *Invest. zool. chil.* **7**, 151–7.

III. Bibliographies

The "Arachnida" section of the *Zoological Record*, produced annually, records all
   publications of the previous year, the list of titles being followed by systematic
   and other analyses. It is an invaluable guide to contemporary research.

Bibliographia Araneorum. P. Bonnet. Three volumes. Published in Toulouse between
   1945 and 1961 this is an immense work of nearly 6,000 pages which contains
   references to everything written about spiders from the earliest times to 1939. The
   list of nearly 8,000 titles is followed by analyses of various topics and finally by
   references to every genus. A supplement is in preparation.

Klassen und Ordnungen des Tierreichs. Edited by H. G. Bronn. The fourth part of
   Volume 5 contains the sections on Arachnida, and each order opens with an
   exceptionally full bibliography, arranged in chronological order. For example,
   of Solifugae there are given 374 titles to 1931 and of Pseudoscorpiones 517 to 1933.

Handbuch der Zoologie. Edited by W. Kükenthal and T. Krumbach. The second half
   of the third volume contains the Chelicerata, and the chapter on each order contains an adequate bibliography, with emphasis on recent work.

Traité de Zoologie. Edited by P.-P. Grassé. The Chelicerata are contained in Tome VI,
   and each order is followed by a sufficient bibliography, which has the advantage of
   being 30 years more up-to-date than the two pre-war works mentioned above.

The Centre International de Documentation Arachnologique produces annually its
   "Liste des Travaux Arachnologiques, parus ou actuellement sous presse". This is
   issued on publication to members of the Centre. Publications bear the limiting
   note, "Sauf les Acariens".

Index Rerum

A

Abnormalities, 57
Acari, 198
Alimentary system, 31
Amber, 320
Amblypygi, 143
Antarctica, 84
Anthracomarti, 173
Arachnid
  appendages, 19
  classification, xi, 14, 103
  diagnosis, 7
  distribution, 12
  evolution, 90, 97
  exoskeleton, 23
  numbers, orders, 8
  orders, size, 13
  somites, 18
Arachnology
  chemical, 282
  economic, 247
  historical, 258
  linguistic, 295
  medical, 288
  practical, 270
Arachnophobia, 316
Araneae, 148
Architarbi, 177
Astigmata, 207
Autecology, 90

B

Behaviour, 71
Bionomics, 60
Blood, 35
Book-lungs, 34

C

Calamistrum, 155
Catalepsis, 73
Chelicerae, 19

Chelicerata, 4
Chitin, 282
Chromosomes, 44
Classifications
  Acari, 209
  Amblypygi, 146
  Arachnida, 103
  Araneae, 164
  Cyphophthalmi, 195
  Opiliones, 188
  Palpigradi, 129
  Pseudoscorpiones, 231
  Ricinulei, 218
  Schizomida, 140
  Scorpiones, 122
  Solifugae, 242
  Uropygi, 186
Claws, 21
Communication, 70
Courtship, 67, 309
Coxal glands, 38
Cribellum, 155
Cucollus, 211

D

Dendrogram, 109
Denticulae, 25
Desiccation, 23
Diplostenoecism, 92
Distribution
  Acari, 208
  Araneae, 159
  Cyphophthalmi, 195
  Opiliones, 186
  Palpigradi, 128
  Pseudoscorpiones, 230
  Ricinulei, 216
  Schizomida, 140
  Scorpiones, 120
  Solifugae, 241
  Uropygi, 135
Drinking, 66

E

Ecdysis, 55
Ecology, 87
Embryology, 44
Ethology, 71
Evolution, 94
Excretion, 38
Eyes, 24

F

Fasting, 66
Fear of arachnids, 316
Feeding, 63
Flagellum, 238
Flash-colours, 72
Flexor reflex, 72
Fluorescence, 287

G

Gastrulation, 46
Genitalia, 37
Geological record, 13, 90, 97
Glands, 41
  epiandrous, 23
  coxal, 38
  odoriferous, 41
  prosomatic, 159
  silk, 41, 156
  venom, 41, 158
Gossamer, 87
Growth, 49
Gynandry, 58

H

Haptopoda, 175
Hearing, 29, 74
Heart, 35
Hormones, 286

I

Instinct, 79
Intersex, 59
Irritant hairs, 63

J

Jumping, 43

K

Kineses, 75
Kustarachnae, 168

L

Larvae, 50
Legs, 21
Longevity, 57
Lyriform organs, 27

M

Malpighian tubes, 33, 41
Mesenteron, 32
Mesostigmata, 202
Migration, 87
Motherhood, 69
Moulting, 55
Muscles, 41

N

Neoteny, 101
Nervous system, 35
Notostigmata, 201
Nymphs, 52

O

Opiliones, 180

P

Palaeontology
  Acari, 209
  Amblypygi, 146
  Araneae, 163
  Opiliones, 187
  Ricinulei, 216
  Scorpiones, 120
  Solifugae, 242
  Uropygi, 136
  Palpigradi, 130
Parthenogenesis, 45
Pectines, 118
Pedicel, 18
Penis, 69
Pheromones, 285
Phoresy, 88
Phylogeny, 94
Preening, 60
Proctodaeum, 33
Prostigmata, 205
Protective colours, 72
Pseudoscorpiones, 220

R

Reflexes, 71
Regeneration, 56
Respiration, 34, 84

Rhythms, 81
Ricinulei, 211

S

Schizomida, 138
Scorpiones, 115
Scrub typhus, 293
Segmentation, 5
Setae, 26
Sex organs, 37, 69
Silk, 157, 283
Solifugae, 233
Spermatophore, 68
Spiders, 148
Spinnerets, 155
Sternum, 18
Stilting, 62
Stridulation, 156, 237
Synecology, 90

T

Tarsal organ, 29

Taxes, 75
Taxonomy, 103
Tetrastigmata, 201
Thought, 77
Trichobothria, 26
Trigonotarbi, 170
Tropisms, 75
Tubercles, 25

U

Uropygi, 132

V

Venom, 284
Vibrotaxis, 76

W

Webs, 303
Wegener's theory, 86

X

X-chromosome, 44

Index Animalium

Where a species is illustrated, the page number is in italic

A

Acarus siro, 208, 251, 252
Aceria fraxinivorus, 206
    sheldoni, 250
Achaearanea tepidariorum, 86
Amblyomma americanum, 255
Amourobius fenestralis, 280
Analges passerinus, 159
Aponomma ecinctum, 203
Araneus diadematus, 80, 258, 280, 289, 310
Architarbus elongatus, 178
    rotundatus, 178
Argas persicus, 204
    reflexus, 254
Argiope bruennichi, 86
Argyroneta aquatica, 162
Atemnus oswaldi, 224
Atypus affinis, 160

B

Belaustium nemorum, 88
Belisarius alleni, 123
    zambeni, 123
Blattisocius tarsalis, 203
Boophilus annulatus, 204
Brevipalpus californicus, 249
Buthotus minax, 82
Buthus occitanus, 116, 294

C

Caddo dentipalpis, 321
Calcitro fisheri, 140
Centruioides sculpturatus, 26
Charon grayi, 145
Cheiridium museorum, 259
Cheiromachus corriaceus, 322
Chelifer cancroides, 56, *221, 229*
Chiracanthium punctorium, 293
Chthonius ischnocheles, 224, 284
Clubiona brevipes, 92
    compta, 92

Clubiona juvenis, 92
Cryptocellus pelaezi, 54, 219
Cryptostemma westermannii, 266
Cupiennius salei, 28

D

Damon variegatus, 147
Demodex folliculorum, 253
    canis, 253
    equi, 253
Dermacentor andersoni, 254, 255
    variabilis, 255
Dermanyssus gallinae, 253
Dermatophagoides pteronyssinus, 208, 256
    scheremetewskyi, 208
Dicranopalpus ramiger, 321
Discotarbus deplanatus, 178
Dolomedes plantarius, 56

E

Ectatostricta davidi, 161
Eotetranychus sexmaculatus, 249
Ero furcata, 156
Eukoenenia mirabilis, 86
Eurypelma californica, 55
Eutypterus fisheri, 95
Evarcha blancardi, 80

F

Filistata hibernalis, 67
    testacea, 45
Francisella tularensis, 255

G

Galeodes arabs, 225, 234
    aranoides, 236, 240
Galumna virginiensis, 208
Garypus birmaticus, 231
Geratarbus lacoei, 178
    minutus, 178
Gigantoscorpio willsi, 121

Glycyphagus domesticus, 252
Gonyleptes janthinus, *187*
    nemastomoides, 187, 322
Graeophrynus anglicus, 146

H
Haemaphysalis spinigera, 255
Hasarius adansoni, 86
Herpyllus blackwalli, 280
Heteropachyloidellus robustus, 286
Heterotarbus ovatus, 178
Holocnemus hispanicus, *154*
Hya heterodonta, *224*
Hyalomma transiens, 254
Hypochilus thorelli, 161
Hypomma bituberculata, 92

I
Isometrus maculatus, 122
Ixodes ricinus, 204, 205, 254
    holocyclus, 255

K
Koenenia mirabilis, *126, 127, 128*, 129
Kustarachne tenuipes, 168

L
Latrodectus mactans, 292
Leiobunum formosum, 286
    longipes, 81, 286
    rotundum, 72
    speciosum, 286
Linyphia triangulasis, 280
Liphistius batuensis, *159*
Loxosceles reclusa, 292
Lycosa amentata, 280
    lenta, 67
    tarentula, 289

M
Mastophora cornigera, 55
Megabunus diadema, 45
Megninia cubitalis, 208
Matta hambletoni, 225, *226*
Metatarbus triangularis, 178
Microbuthus pusillus, 122
Moniezia expansa, 208
Mygale lasiodora, 289

N
Nemastoma bimaculatum, *184*
    denticulatum, 321
Neobisium imperfectum, *224*

Notiomaso australis, 85
Nyctalops crassicaudatus, *139, 140*

O
Obisium simile, *223*
Odiellus gallicus, 58
Ophionyssus hirsti, 203
Opilio ovalis, 321
    parietinus, *184*
Oppia minuta, 208
    ornata, 208
Ostearius melanopygius, 86
Oribata castanea, 159
Ornithodorus lahorensis, 254
    moubata, 255
    savignyi, 204
Ornithonyssus bursa, 253
Otobius megnini, 204, 255
Otodectes cynotis, 208, 256
Oxypleurites neglectus, 251

P
Palaeophonus nuncius, 13, 120
Pandinus gambiensis, 122
    imperator, 122
Panonychus citri, 249
    ulmi, 249
Pardosa lapidicina, 286
Penthaleus belli, 85
Phalangiotarbus subovalis, 178
Phalangium opilio, 45, 56, 81, 182, 259
Philodromus fallax, 72
Pholcus phalangioides, 74
Phyllocoptes masseei, 251
Pirata piraticus, 79
Plesiosiro madeleyi, *175, 176*
Polyochera alticeps, 217
    glabra, 217
    punctulata, 217
Polyphagotarsonemus latus, 250
Protacarus crani, 209
Protosolpuga carbonaria, 242, *243*
Pseudomonas hydrophilus, 203
Ptilonyssus hirsti, 203
Pyemotes ventricosus, 256

R
Rhipicephalus evertus, 203
    sanguineus, 255
Rhizoglyphus echinopus, 251
    parasiticus, 252
Ricinoides afzelii, *216*, 219

Ricinoides crassipalpe, 212, 213, 214, 216
    sjostedti, 219
    westermanni, 215
Rickettsia tsutsugamushi, 207
Rubrius subfasciatus, 85

S

Salticus scenicus, 280
Sarcoptes scabiei, 208, 256
Schizocosa crassipalpes, 285
Scirus longirostris, 206
Segestria senoculata, 72
Siro rubens, 58, *192*
Solpuga monteiroi, 240
Steatoda dipunctata, 280
    lepida, 305
Stegophrynus dammermani, 144
Stenotarsonemus pallidus, 250
Sternarthron zitteli, 130
Sternostoma trachaecolum, 203
Subacon bachofeni, 322
Symphytognatha globosa, 162
Synageles venator, 92

T

Tachypleus (Limulus) gigas, 248

Tarentula cubensis, 289
    hispana, 289
Tegenaria atrica, 151, 281
    derhami, 280
    parietina, 45
Telema tenella, 161
Tetranychus cinnabarinus, 207
    urticae, 249
Thelyphonus insularis, 133
Tibellus maritimus, 92
Tidarren fordum, 63
Titanoeca albimaculata, 305
Tityus eogenus, 121
Trigonomartus pustulans, 171
Trombicula autumnalis, 207, 257
Tyrophagus longior, 251, 252
    putrescentiae, 251, 252

X

Xysticus cristatus, 149, 280

Z

Zercoseius ometes, 202
Zora spinimana, 156
Zygiella atrica, 81